SPRINGER
LABORATORY

A. Radbruch (Ed.)

Flow Cytometry and Cell Sorting

With 58 Figures and 6 Tables

Springer-Verlag
Berlin Heidelberg New York
London Paris Tokyo
Hong Kong Barcelona
Budapest

Prof. Dr. Andreas Radbruch
Institut für Genetik, Universität Köln
Weyertal 121, 5000 Köln 41, FRG

QH
585
.5
.F56
F56X
1992

ISBN 3-540-55594-3 Springer Verlag Berlin Heidelberg New York
ISBN 0-387-55594-3 Springer-Verlag New York Berlin Heidelberg

The use of general descriptive names, registered names, trademarks, etc. in this publication does not imply, even in the absence of a specific statement, that such names are exempt from the relevant protective laws and regulations and therefore free for general use.

Product liability: The publishers cannot guarantee the accuracy of any information about dosage and application contained in this book. In every individual case the user must check such information by consulting the relevant literature.

Cover Design: Struve & Partner, Heidelberg
Data conversion: Struve & Partner, Heidelberg

66/3145 – 5 4 3 2 1 0 – Printed on acid-free paper

Contents

Part IV Cellular Activation and Biochemistry

Part V Cell Sorting

Part VI Safety

List of Contributors

Assenmacher, Mario
Institut für Genetik
Universität zu Köln
Weyertal 121
5000 Cologne 41
FRG

Brown, Dr. Spencer C.
Service de Cytométrie
Institut des Sciences Végétales
CNRS
91198 Gif-sur-Yvette
France

Esser, Dr. Charlotte
Institut für Umwelthygiene
Auf 'm Hennekamp 50
4000 Düsseldorf 1
FRG

Fantes, Dr. Judith A.
MRC Human Genetics Unit
Western General Hospital
Crewe Road
Edinburgh EH4 2XU
UK

Galbraith, Dr. David W.
Department of Plant Sciences
University of Arizona
Tucson, Arizona 85721
USA

Göttlinger, Christoph
Institut für Genetik
Universität zu Köln
Weyertal 121
5000 Cologne 41
FRG

Green, Dr. Daryll K.
MRC Human Genetics Unit
Western General Hospital
Crewe Road
Edinburgh EH4 2XU
UK

Herzenberg, Prof. Dr. Leonard A.
Stanford University School
of Medicine
Department of Genetics
Stanford, CA 94305
USA

Hinnisdaels, Dr. Steph
Plant Genetics
Institute for Molecular Biology
Paardenstraat 65
1640 St. Genesius Rode
Belgium

Irlenbusch, Sigrid
Institut für Genetik
Universität zu Köln
Weyertal 121
5000 Cologne 41
FRG

Jung, Steffen
Institut für Genetik
Universität zu Köln
Weyertal 121
5000 Cologne 41
FRG

Kubbies, Dr. Manfred
Boehringer Mannheim GmbH
Flow Cytometry Laboratory
Nonnenwald 2
8122 Penzberg
FRG

Kato, Dr. Kimitaka
Institut für Genetik
Universität zu Köln
Weyertal 121
5000 Cologne 41
FRG

Marie, Dr. Dominique
Service de Cytométrie
Institut des Sciences Végétales
CNRS
91198 Gif-sur-Yvette
France

Mechtold, Birgit
Institut für Genetik
Universität zu Köln
Weyertal 121
5000 Cologne 41
FRG

Meyer, Klaus L.
Sinnersdorfer Feld 148
5024 Pulheim 4
FRG

Miltenyi, Stefan
Miltenyi Biotec GmbH
Friedrich-Ebert-Str. 68
5060 Bergisch Gladbach 1
FRG

Müller, Dr. Werner
Institut für Genetik
Universität zu Köln
Weyertal 121
5000 Cologne 41
FRG

Murphy, Dr. Robert F.
Department of Biological Science
Carnegie Mellon University
4400 Fifth Avenue
Pittsburgh, PA 15213
USA

Otto, Dr. Friedrich J.
Fachklinik Hornheide
Abteilung Experimentale Onkologie
Dorbaumstr. 300
4400 Münster-Handorf
FRG

Pflüger, Dr. Eckhard
Miltenyi Biotech GmbH
Friedrich-Ebert-Str. 68
5060 Bergisch Gladbach 1
FRG

Radbruch, Prof. Dr. Andreas
Institut für Genetik
Universität zu Köln
Weyertal 121
5000 Cologne 41
FRG

Recktenwald, Dr. Diether
Becton-Dickinson
Immunocytometry Systems
2350 Qume Drive
San José, CA 95131 1893
USA

Rothe, Dr. Gregor
Klinikum der Universität
Regensburg
Franz-Josef-Strauß-Allee 11
Postfach 10 06 62
8400 Regensburg
FRG

Valet, Prof. Dr. Günter
Max-Planck-Institut für Biochemie
8033 Martinsried bei München
FRG

Veuskens, Dr. Jacky
Plant Genetics
Institute for Molecular Biology
Paardenstraat 65
1640 St. Genesius Rode
Belgium

Weichel, Dr. Walter
Bayer AG im Zentralbereich
Forschung
Forschung Biotechnologie
Gebäude Q18
5090 Leverkusen
FRG

Introduction

L.A. HERZENBERG

Initially (in 1968), the fluorescence-activated cell sorter (FACS) was an instrument run for and by experts. With some tweaking and twisting, the FACS engineering prototype produced at Stanford by engineers Bill Bonner, Dick Sweet and Russ Hulett working with me could be convinced to do flow cytometric analysis and to physically sort cells without compromising cell viability. Aseptic sorting was added early on, so that sorted cells could be grown in culture or transferred into irradiated mice or rabbits. Yes, some of the first adoptive transfer experiments with FACS-sorted cells were performed with rabbits (in collaboration with Dr. Patricia Jones and Dr. John Cebra) because fluorescent antibody reagents detecting the then more extensively known immunoglobulin (Ig) heavy and light chain allotypes of the rabbit were readily available for staining, sorting and subsequent fluorescent microscope analysis to distinguish the origins of the transferred cells.

The first commercial FACS instrument, produced in collaboration with us by an engineering group under Bernard Shoor and David Capellaro at Becton Dickinson, Inc. (ca. 1973), replaced the engineering prototype we had been using in our laboratory. The next three FACS instruments, also produced by Becton Dickinson, Inc., landed with John Wunderlich and Susan Sharrow at the NCI in Bethesda, Mel Greaves and Avrion Mitchison in London, and Klaus Rajewsky (later joined by Andreas Radbruch) in Cologne. From this auspicious beginning, the number of FACS instruments and the attendant technology expanded over a period of years to its current proportions: several thousand instruments produced by a number of manufacturers, serving a wide variety of skilled and novice users with interests that range from basic immunology and molecular biology to clinical medicine and oceanography. The introduction of monoclonal antibodies as highly specific staining reagents that could be produced indefinitely in large quantities and readily labeled with a variety of fluorochromes overcame the last barrier to the apparently limitless applications of flow cytometry in all areas of medicine, cell biology and the ancillary fields of modern biology.

The impetus for building the first FACS was generated (ca. 1965) by frustration with the quantitative and qualitative limitations of fluorescence microscopy. Basically, we could see and count cells tagged with fluorescent-labeled antibodies; however, we had no way to separate them and define their functions (as we do now with FACS-sorted populations or with clones generated from individual sorted cells). Spleen cells bearing surface Ig molecules, for example, were readily visualized and seemed likely candidates for precursors of antibody producing cells; however, the best that could be done prior to the development of the FACS was to correlate the frequency of these cells obtained after immunization with the frequency of cells that produced antibodies. Similarly, although thymic-derived (T) cells could be clearly identified by microscopy, characterization of the development and function(s) of these cells only became possible after FACS and monoclonal antibody technology introduced the potential for using many different fluorescent markers to analyze and sort these cells for functional studies.

Not surprisingly, the necessity to identify, count and sort viable fluorescent-tagged cells that mothered the invention of the FACS some 25 years ago has now spawned an entire field devoted to finding new and better ways to use flow cytometry to characterize and isolate cells. For example, much better fluorochromes were needed to stain cells brighter and with higher specificity, i.e. greater signal to noise ratios. Sea weeds and algae were extracted to provide these marvelous „antenna molecules" known as phycobiliproteins.

Similarly, intracellular reporter molecules were needed to allow FACS measurement of the individual cell distribution of gene transcription and/ or expression. For these purposes, we adapted an old fluorogenic system by using the E.coli *lacZ* gene (which codes for the enzyme β-galactosidase) as a reporter gene to ligate to promoters, enhancers and other mammalian DNA regulatory sequences. To detect the expression of *lacZ* in these labeled cells, we used hypotonic treatment to introduce a fluorogenic β-galactosidase substrate (fluorescein phycobiliproteins-galactoside) into the cells transfected with the constructs and then „incubated" the substrate-loaded cells at temperatures below the membrane lipid „freezing point" in order to retain the cleavage product (fluorescein). With this methodology, we succeed in using the FACS to select and clone cells expressing *lacZ* under the control of various genetic regulatory sequences and to measure changes in the expression of this reporter gene under different growth and/or stimulatory conditions.

The reporter gene methodology, of course, was built on earlier methods developed to sort and clone rare cells, e.g., mutants and transfectants. Rajewsky, we (David Parks, Tom Kipps) and many others in Cologne, Stanford and elsewhere worked on these methods, which currently enable the sorting of cells present at frequencies of 1 per 10 million or lower and the direct cloning of the sorted cells to generate cell lines expressing the phenotype of the sorted cells. Thus the FACS has become a major instrument for molecular and somatic cell genetics studies.

Multiparameter analytic and sorting methods bring added dimension to these *in vitro* studies in addition to being central to FACS studies characterizing the development or functions of subsets of mammalian cells. Much of this technology was developed in studies of the immune system, where the measurements of the quantitative expression of three or four cell surface determinants are often required to resolve functional or developmental subsets/lineages. Immunological studies also provide perhaps the most widely known FACS use of the FACS: the determination of T cell subset frequencies used to monitor disease progression in HIV-infected individuals.

Dr. Radbruch and the authors of this volume provide readers with the „how to" of these and many other modern methods for flow cytometry and fluorescence activated cell sorting. I recommend the reading and trying of these methods, and challenge readers to develop novel and better methods to exploit this flexible and powerful technology.

Part I Flow Cytometry

1 Operation of a Flow Cytometer

C. GÖTTLINGER, B. MECHTOLD, and A. RADBRUCH

1.1 Background

Biomedical research, on the way from the organism to the molecule and back again, requires powerful tools to analyze the functional status of the individual cell, the unit of organization of life. Cells were detected by microscopy, and microscopy, in combination with powerful staining technologies, continues to be the main instrument for obtaining direct information on their state of activation, proliferation, and differentiation. The drawback of microscopy is that the data generated are mainly „visual impressions“ and not exact numbers. Today, systems are available that allow quantification of the light intensity of microscopic objects, but these systems are still too slow to allow analysis of enough objects to obtain good statistics. One could say that microscopy generates too much information in cases where information only on the amount of a particular stain per cell is desired.

The idea of rapidly and precisely measuring high numbers of stained cells was realized between 1965 and 1970 by combining electronic light measurement and the concept of letting a flow of small cells pass the microscope's objective rather than moving the large microscope over the cells [1,2]. This basic concept of flow cytometry is realized in a variety of commercially available instruments, including Coulter's PROFILE and EPICS, Becton-Dickinson's FACScan, FACStrak, FACSort, and FACStar, Ortho's CYTORON, and Partec's PAS machines. For the hobby engineer, Shapiro has described a build-it-yourself cytometer, the CYTOPUP and CYTOMUTT [2].

The application range of flow cytometry has grown ever since its original introduction, mainly due to the development of powerful staining technologies that make use of the high sensitivity of fluorescence, the principle of chromogenicity, i.e. color shifts caused by biochemical reactions of the dye, and monoclonal antibodies to stain cells for specific proteins. Methods are available to quantify DNA content very precisely (Chap. 7), determine the proliferative history of a cell in vivo (Chap. 8) and in vitro (Chap. 9), to measure ion fluxes correlated to the cell's physiology (Chaps. 10–12) and enzymatic activities, and to discriminate among cells in complex mixtures by immunofluorescence (Chaps. 2–4).

Here, we describe the principal components of flow cytometers, the fluid system that transports the cells across the microscopic field, the optics of illumination and detection, and the electronics for light collection, data managment and control. Basic setup and calibration routines are given for a typical fixed aligned flow cytometer and a nonfixed aligned flow-in-air cytometer. Procedures for other machines may vary slightly. In any case, the manufacturer's instructions should be observed.

1.1.1 Fluidics System

The flow system is shown schematically in Figure 1. The stained cells are applied to the cytometer in a „sample tube." From the sample tube the cells are transported by air pressure into the flow chamber, which can be a cuvette directly observed by the microscope or a nozzle that injects the cells into the air and through an area of microscopic observation (flow-in-air system).

Essential for the success of flow cytometry is hydrodynamic focusing [3], by which the cells are individualized and positioned at the observation point with an accuracy of better than 1 μm. For hydrodynamic focusing, the cell suspension (sample fluid) is injected into a particle-free „sheath fluid" in a large-diameter tube which flows in a small-diameter cuvette or a nozzle with a typical orifice diameter of 70 μm. The diameter of the sample flow

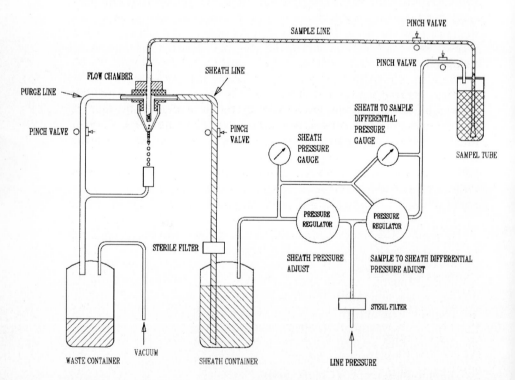

Figure 1. Diagram of the fluid system of a flow cytometer with differential pressure sample injection system

is reduced from about 200 µm at the injection point to about 10 µm in the cuvette or nozzle orifice. It is clear that the relationship between sample and sheath pressure is a critical parameter for optical resolution as well as the flow rate of cells. Increasing the „sample pressure" to increase the flow rate also increases the diameter of the sample stream and thus may lower the quality of the optical analysis.

Usually, sheath and sample flow are controlled by two pressure regulators (Fig. 1). Sheath pressure is established once and not changed over months. A new setup is required only if the diameter of flow is changed, or if the sheath in-line sterility filter has been exchanged. In flow cytometers with fixed setup (e.g., FACScan) sheath pressure is also fixed. Sample pressure can be changed either gradually (flow-in-air systems) or stepwise (FACScan). In most machines additional lines are connected to the flow chamber to allow removal of trapped gas bubbles and debris or to clean the chamber.

1.1.2 Optical System

Except for the Partec machines, which have an optical geometry closely resembling that of a microscope, flow cytometers have a three-dimensional orthogonal optical geometry, in that the liquid stream, illuminating light beam, and microscopic axis are all perpendicular to each other (Fig. 2).

The optical system consists of three parts:

Illumination Optics. In principle, the light source for flow cytometry can be either a conventional lamp, (e.g., a high-pressure mercury lamp), or a laser, generally an argon ion laser. If the light source cannot be directed directly at the liquid stream, prisms and mirrors are required to direct the light. If the light source does not provide monochromatic light, filters must be used. Several spherical or cylindrical lenses are required to focus the illuminating light onto the point of microscopic observation. For systems using a laser light source, additional optics can be used to spread the laser beam first, in order to protect prisms, mirrors, and filters.

Forward Scatter Collection Optics. A microscope with low numerical aperture observes the liquid stream from opposite the illuminating light to collect light scattered by particles in the liquid stream in the range of 2°–20° off the axis of the illuminating light.

Fluorescence and Side Scatter Collection Optics. A second microscope observes the cells from the side, i.e., perpendicular to the illuminating light axis as well as the liquid stream. This objective has a long working distance and high numerical aperture, typically 0.6–0.65 for flow-in-air cytometers and 1.2 for cuvette cytometers. This is one of the reasons why cuvette cytometers usually give better performance in terms of sensitivity. The advantage of flow-in-air systems is their better sorting capability (see Chap. 16).

A diagram of the optical configuration of a flow-in-air sorter equipped for detection of four optical parameters per cell is presented in Figure 2. The

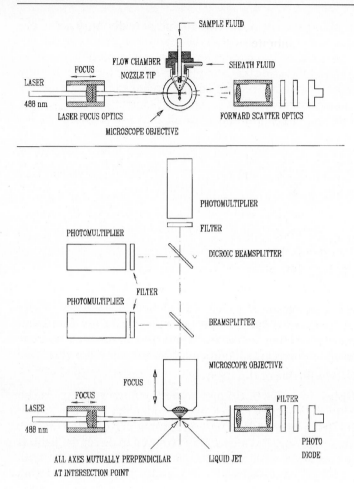

Figure 2. Typical optical configuration of a laser-based flow cytometer. In this design the flow chamber is formed by a special nozzle holder with a nozzle tip. The detection area is located approximately 0.3 mm down the lower edge of the nozzle tip. Short light pulses of scattered laser light or fluorescence emission are produced each time a cell or particle passes the focused laser beam. The duration of light pulses is typically in the range of 3–20 µs depending of flow velocity, laser focus spot, and particle size. Scattered laser light from the cells is collected with above-described optics, and the light pulses are converted to electrical current pulses with appropriate optical sensors. The unscattered laser light passing the liquid jet is blocked by an obscuration bar in front of the forward scatter optics. In front of the fluorescence optics an obscuration bar is also neccesary to block unscattered light reflected from the surface of the round liquid jet. Other nonsorting instruments use a rectangular flow in a rectangular quarz cell with high-quality optical surfaces. With such a design, coupling of excitation light to the flow and outcoupling of fluorescence light is very efficient. Reflecting of unscattered excitation to the fluorescence collecting optics is minimized so that no obscuration bar is necessary in front of this optical arrangement. The mixture of the strong, scattered laser light and dim, fluorescent emission from different dyes, collected by the microscope objective is separated by optical filters. After passing the filters the fluorescence light pulses from the cells are converted to pulses of electrical current via photomultipliers tubes

orthogonal geometry of illumination and the cell path and detection make it difficult to align and calibrate such a system.

1.1.3 Electronics

A simplified block diagram of electronic signal detection and processing is shown in Figure 3. The various elements are described below.

Photodetectors. Measuring light in flow requires light sensors that convert light signals into electrical signals. For detection of forward light scatter a semiconductor photodiode is sufficient. The dim fluorescence and side scatter signals are detected by photomultiplier tubes (PMT) in which photons elicit a cascade of electrons. The amplification is controlled by the voltage applied to the PMT to sensitize it. This voltage ranges from 300 to 1000 V. Depending on the voltage, the amplification can vary between 10^3 and 10^8. The relationship between supply voltage (V) and gain (G) of amplification is characteristic for a given PMT [4]:

$G = K$ x V^α, with α = 6–9 (depending on type of PMT) and K = constant.

Caution: Never expose PMTs to daylight when they are connected to the power supply! Operate PMTs only with an appropriate filter in front to block light from laser or other light sources. The electronic signals from the PMT are taken up by special preamplifiers which convert the current into voltage signals, amplify them, filter out noise, and often provide baseline restoration.

Electronic Fluorescence Compensation. Several popular fluorescent dyes show that spectral overlap of fluorescence emission and complete optical

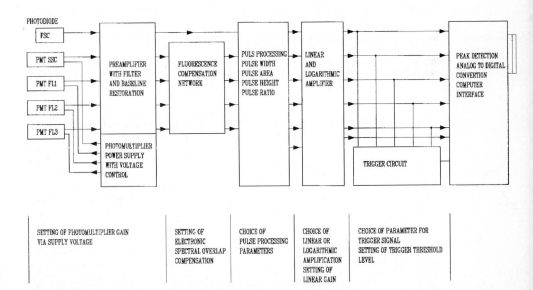

Figure 3. Block diagram of the electronics of a flow cytometer

separation by filters is not possible, at least not if the same wavelength of light is used for excitation, and sensitivity of staining must be retained. An example is the popular combination of the two dyes fluorescein isothiocyanate (FITC) and phycoerythrin (PE) excited by 488 nm argon laser light for immunofluorescence. The 570 nm PE fluorescence detector picks up some fluorescence from FITC, and the detector for 530 nm FITC fluorescence picks up some PE fluorescence. To compensate for this cross-talk, a fraction of FITC sensor output signal is subtracted from the output of the PE output signal and vice versa. This subtraction is done by the electronic compensation network. The rate of spillover is determined by analysis of cells stained exclusively with one or the other fluorochrome (see setup procedures). Note that changing the gain of one sensor, for example, altering the PMT supply voltage, requires new calibration of compensation [5]. For the future design of flow cytometers, compensation should be performed in the final data analysis.

Pulse Processing. For laser-based flow cytometers with an area of illumination larger than the size of particles the height of the analog signal pulse is converted into a digital signal. In the case of pulses from single round particles, such as cells, the peak is proportional to the area and thus can be taken as a measure for the intensity of light. The height of the peak can also be determined for weak and noisy signals. Precise determination of the area and width of signal pulses requires strong signals, but it gives information on the shape of the pulse, and it is linear over a wider range of fluorescence if particles larger than or approaching the size of the illuminated area are analyzed. Pulse area and pulse width are most useful for discrimination of G_2/M phase cells from cell doublets, which is critical for cell cycle analysis [6,8]. Determination of pulse width for forward scatter allows quantitative measurement of cell size if the scale has been calibrated with beads of known size, and the laser focus is smaller than cell size [6–8].

Linear and Logarithmic Amplifiers. Flow cytometers are usually equiped with both linear and logarithmic amplifiers. Linear amplification is generally used for parameters with small dynamic range, for example, DNA content. The gain setting of linear amplification usually varies between one- and 32-fold. Immunofluorescence has a wider dynamic range, and differences in intensity of more than 1000-fold are not rare. In logarithmic amplification, this dynamic range can be displayed on one scale, and the resolution is still very good for low signals. Also, since the distribution of different intensity but same CV give histograms of the same width on a logarithmic scale, it is easier to visualize populations with large CV (see Chap. 3). For the lower and upper ends, logarithmic amplifiers show individual deviations from the ideal linear to logarithmic conversion. These deviations must be determined for each amplifier before comparisons of relative fluorescences can be made. The calibration curves can be obtained rather simply [12,13].

Trigger Circuit. To separate signals from electronic noise or debris, the signal of one parameter (scatter or fluorescence) is used for triggering the processing of all correlated signals by the electronic. The signal of the parameter used for triggering must exceed a threshold level, set by the operator, to permit further signal processing. Setting the trigger threshold can be regarded as setting a „live gate" (see Chap. 3), allowing the exclusion of debris in the case of cell cycle analysis and flow karyotyping, triggering on DNA fluorescence, or erythrocytes in the case of immunofluorescence, using forward scatter as trigger.

Analog/Digital Conversion. Further processing of the signals from all parameters is started if the input parameter of the trigger circuit exceeds the trigger threshold. First, the peak of each analog pulse is determined, held, and converted into a digital number which is transmitted to a computer for storage and data evaluation.

1.2 Material

The sheath fluid should be a physiological fluid with about the same optical characteristics as the sample fluid. The sheath fluid should be filtered to remove debris and to sterilize it (0.22-µm filters). As an additional precaution for sterile sorting, 0.03% sodium azide (NaN_3) can be added; however, this is controversial since the physiology of the cell may be altered although sodium azide acts reversibly and can be washed out of the sorted cells. For mammalian cells use 0.9% NaCl in H_2O. **Sheath fluid**

The optical filters for the various dyes are listed in Table 1. **Optical filters**

The sample filter should prevent anything with a diameter larger than the diameter of the nozzle from entering the nozzle and blocking it. The filter is inserted at the end of the sample line. Sample filters can be purchased or self-made (see Sect. 1.4). **Sample filters**

- A tissue-culture compatible detergent, diluted in water and filtered through 0.22 µm filters, for example, 0.1% 7X (Flow). **Cleaning solutions**
- Sodium hypochloric acid (Chlorix), which is diluted just before use to 5% with water. Chlorix is very efficient in cleaning and sterilizing the cell path. In water it is rapidly degraded into sodium chloride and water.

- Chicken red blood cells (CRBCs), fixed with glutar aldehyde and dispersed by sonification. **Calibration standards**
- Stained and unstained plastic beads or latex beads, for example, Calibrites (Polysciences or Flow Cytometry Standards Corporation).
- Fixed cells, unstained and stained with dyes that require compensation if used in combination.

Nylon gaze (30, 45, or 60 µm pore size) (Erbslöh, Düsseldorf, or Schweizerische Seidengazefabrik AG, Zürich). **Miscellaneous**

Table 1. Some fluorochroms for protein staining

Fluorescence dye	Abbr.	Excitation Peak nm	Laser line and type nm	Emission Peak nm	Emission[a] Filter	Emission[b] Filter
Fluorescein	FITC	495	488 Ar	520	DF530/30	DF530/30
R-Phycoerythrin	R-PE	565,545,480	488 Ar	575	DF585/42	DF575/26
Texas Red	TX	595	595 Dye R6G	620		DF630/22
Tandem PE-TX		see PE	488 Ar	613	DF625/30	
Allophycocyanin	APC	650	632 HeNe, 595	660		DF660/20
Tandem PE-Cy5		see PE	488 Ar	670	DF675/30	
Per. chlorophyll[c]	PerCP	470	488 Ar	680	DF680/20	

[a] Emission filters for single laser (488 nm) excitation. In addition, for up to four color immunofluorescence measurements, three dicroic mirrors are required. The fluorescence emission from the different dyes is split first by a 600 nm shortpass dicroic mirror in a green/yellow (FITC/PE) and a red (red 613/red 670) wavelength range. A DM 560-nm shortpass dicroic separates FITC from PE fluorescence and a 640-nm longpass dicroic mirror separates the emission from the different red dyes. (Note: At the moment no commercial instrument allows electronic spillover compensation for four colors excited with one laser (maximum three colors), but we are certain that this will be possible in future after some changes or upgrades in the compensation network of the instruments.)

[b] Emission filters for dual laser excitation with 488-nm argon and 595-nm dye laser. The two laser focus spots on the liquid jet are separated physically by a small distance (typically 0.1 mm). The fluorescence emission from one laser spot (TX/APC) is deflected at a right angle toward the optical sensors by a small mirror. Separation of FITC/PE and TX/APC emission is performed by two dicroic mirrors as described above.

[c] Peridinin chlorophyll-a protein complexes [14].

1.3 Methods

1.3.1 Setting Up a Flow Cytometer with Fixed Alignment

Flow cytometers with fixed alignment are very widely distributed. They offer rather easy operation, reproducible results, and high sensitivity. The cells are illuminated and measured in a cuvette with fixed optical alignment. State-of-the-art instruments are equipped with a small argon ion laser (15–20 mW; Spectra-Physics), measure five optical parameters, and can store and restore the instrument settings. The setup procedure includes checking the sheath fluid reservoir, switching the instrument and the computer on, downloading the instruments settings, and checking the performance with standards. This is exemplified below for the FACScan flow cytometer.

1. Check sheath container (left) and waste container (right). Fill up sheath and empty waste if necessary. Add detergent to the waste container for decontamination. Caution: Do not fill the sheath container beyond the upper mark! Check tubings from and to fluid containers to make sure that they allow free flow.

2. Check that the vent valve toggle switch between the fluid containers is in the upward position.

3. Turn the main power switch on. This starts the laser, air compression, vacuum, and the electronics. After about 5 min, required for warming up, the NOT READY sign will change to STANDBY if no sample tube is inserted yet, or to READY if a sample tube is inserted and the fluidics control switch is set to RUN.

4. Turn on the periphery of the computer, i.e., screen, hard disk, printer etc., and then the computer itself, or all at once if a main switch is installed. The computer loads the FACScan software automatically.

5. Switch the fluidics main control from STANDBY to FILL for approximately 30 s.

6. Check the sheath filter for trapped air. If air bubbles are visible, blow them out through the screw cap on top of the filter by slightly unscrewing the cap until the air is followed by saline. Tighten the cap again.

7. If air is trapped in the flow chamber (cuvette), switch the fluidics control to DRAIN, watching the chamber. When the sheath fluid retracts from the chamber after a few seconds, turn the fluidics control to FILL, and allow the fluid to fill the chamber again, letting the air escape to the top. Finally, switch the fluidics control to STANDBY again. Note: This procedure is also advisable for slight clogging of the chamber, using the surface of the retracting sheath to clean the walls of the chamber mechanically.

8. Set switch for flow rate to LOW.

9. Start aquisition software.

10. Load setting of instrument for your experiment from earlier experiments or standard files. Initially, the settings must be determined using standard particles, according to the manufacturer's advice. For a variety of reasons, this may not yield sufficient results. In this case, in addition to the standard particles use cells that are unstained and stained exclusively with the dyes in question for fine-tuning of sensitivity and compensation. This is described below.

Since alignment of a FACScan is stable, and instrument settings can be stored and recalled for later analyses, determination of optimal threshold, gain, and compensation must be performed only initially or if the cytometer has been repaired. Different settings may be necessary for various combinations of dyes and for different cell types or viable versus fixed cells.

Determination of FACScan settings for viable or formaldehyde-fixed lymphocytes with stained with FITC- or PE-conjugated antibodies, using Calibrite standard beads and AUTOCOMP software is described in [11].

A procedure for three-color FITC/PE/peridinin chlorophyll (see Chap. 4) is described in [10]. Here we describe a simple and general procedure for manual determination of the instrument settings using unstained and stained cells, in this case for three-color immunofluorescence of viable or formaldehyde-fixed lymphocytes with FITC/Pe red 670 (a tandem conjugate of PE and Cy5). The cells can be stained with any antibody conjugates representing the three colors: actually even the same antibody could be taken, conjugated to various fluorochromes. The cells are stained as independent samples, for one color each. Cells of the various samples are then used pure or mixed as indicated below.

1. Set all compensation levels to zero.
2. Select FSC as parameter for threshold trigger. Set trigger level to about 50.
3. Run unstained cells and adjust gains of forward and side scatter until obtaining a dot plot similar to that shown in Figure 4, with the mean of lymphocyte FSC somewhere between channels 300 and 500 (with 1024 channels full scale).
4. Adjust trigger level to exclude debris but to include all lymphocytes.
5. Set a live gate around the population of lymphocytes.
6. While running the sample of unstained cells, adjust voltages of fluorescence detection PMTs so as to display all cells on the left side of the scale, with a mean fluorescence of about 120 channels. Start at 200 V.
7. Change the sample to a mixture of unstained cells and cells stained with FITC. Use F1/F2 (FITC/PE) dot plot display. Increase FL2-%FL1 compensation until the FITC-stained cells appear „negative" in the PE PMT, i.e., have a mean FL2 fluorescence of about 120, as the unstained cells.
8. Change the sample to a mixture of unstained and PE-stained cells. Increase FL1-%FL2 compensation until the stained cells appear „negative" in the FITC PMT, i.e., show a mean FL1 fluorescence of about channel 120.
9. Change dot plot parameters to FL2/FL3 (PE/red 670). Increase FL3-%FL2 compensation until the PE-stained cells appear „negative" for red 670, i.e., have a mean FL3 fluorescence of about channel 120.
10. Change the sample to a mixture of unstained and red 670 stained cells. Increase FL2-FL3% compensation until the red 670 stained cells appear „negative" for PE, i.e., show a mean FL2 fluorescence of about channel 120.
11. Change the sample to a mixture of unstained cells and cells stained for all three fluorochromes. Aquire some data and store them as a „master file" at an easily accessible place with a characteristic name. From this file the instrument settings can be retrieved routinely.

CG2069201

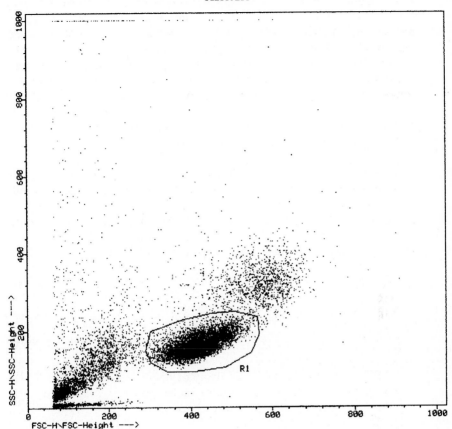

Figure 4. Forward against side scatter dot plot of peripheral blood after removing most of the erythrocytes by Ficoll. The lymphocyte population is marked

When the last sample has been measured:

Shutdown of the FACScan

1. Flush the system with 5% sodium hypochloride (Chlorix) solution in water for 5 minutes.
2. Insert a sample tube with 1–3 ml of 0.1 % solution of 7X (Flow) to prevent drying of the sample capillary tube and clogging with salt crystals. Allow this to run for 1–3 min at HIGH flow rate.
3. Set the fluidics control to STANDBY.
4. Turn off the computer and its periphery. Turn FACScan main switch off.

1.3.2 Setting Up a Free-Flow-in-Air Cytometer (FACStar)

Alignment and calibration procedures for free-flow-in-air cytometers belong to the more demanding technical challenges in flow cytometry and require some experience. One could actually ask why such machines are still in use. The answer is that they offer the best flow sorting capabilities (see Chap. 16–20), and due to their open architecture, which makes them difficult to set up, they are more versatile than machines with fixed alignment. The use of other or additional light sources allows the use of dyes which cannot be used with 488-nm argon lasers (e.g., Chap. 9,10,13). Setting up a flow-in-air cytometer can be divided into basic and routine setup procedures. Basic alignment is required when the optical arrangement is changed. This is usually done by an experienced operator. Until the next change, the alignment then requires only minor calibration, which can be carried out by less experienced persons using standard particles and cells.

The aim of basic alignment is as follows:

1. Aligning the first laser beam (usually a 488-nm argon laser; for example, Spectra-Physics, Coherent) in line with the axis of the laser focus optics and forward scatter detection optics. If an optional second laser is used, which could be a UV laser, dye laser, or helium/neon laser (Spindler and Hoyer, Göttingen), it must be aligned to cross the liquid jet about 0.1–0.2 mm below the intersection point of the first laser.
2. Initial setup of sheath pressure, nozzle position, and stream viewing optics.
3. Adjusting the position of the fluorescence microscope for maximum sensitivity, i.e., aligning the optical axis of the microscope vertically to both the axis of the laser and the axis of the liquid jet.

The procedure for basic alignment varies from machine to machine and is usually described rather well in the manufacturer's manual.

Basics 1. Check sheath fluid and waste containers. Fill up the sheath container (same as for FACScan) and empty the waste container if necessary. Do not fill the sheath container too much: be sure that the input line for compressed air does not dip into the sheath fluid. Add some detergent to the waste container for decontamination.
2. Install appropriate filters for the desired dyes (a list of filters for common dye/laser combinations is given in the table above).

Starting the laser 1. Turn on laser cooling.
2. Check that the laser beam is blocked and move the nozzle tip up so that it cannot be hit later by the laser beam.
3. Switch laser power supply to CURRENT mode and set plasma current to MINIMUM.
4. Switch on the laser. After a few seconds the laser starts.

1. Turn on the power switch of the electronics console. The last settings are loaded automatically.   **Starting the electronics**
2. Switch on the computer (first the periphery then the computer, or all together with main switch). After some time, the main menu appears on the screen.
3. Start acquisition software. The last settings stored in the computer are transferred to the console.

The aim of aligning the nozzle is to align the jet of hydrodynamically focused cells vertically to both, the laser beam and the axis of fluorescence detection. The cells should pass the center of the focused laser beam and the focus of the fluorescence microscope. The point of intersection is usually about 0.3 mm downstream from the tip of the nozzle. At this distance from the nozzle tip surface irregularity of the liquid jet is low; this prevents modulation of the laser beam by such irregularities, which can severely disturb the measurement of scatter.   **Alignment of the nozzle**

1. Turn on the compressed air and the vacuum, with fluidics main control switch still off.
2. Remove the nozzle tip and clean it with a tissue-culture compatible detergent (e.g., 7X, Flow) either by flushing with a syringe or in an ultrasonic bath (30 s; ceramic nozzles only). Put the nozzle tip back on the nozzle holder.
3. Position the nozzle over the center of the microscope (approximately 3 mm, Z control).
4. Turn fluidics control to FILL position for a few seconds and control whether the liquid jet runs down vertically into the waste collector. Adjust if necessary.
5. Check nozzle for trapped air bubbles and remove by switching fluidics control between FILL and SHEATH.
6. Unblock the laser beam and adjust liquid stream to intersect with the laser beam (Y control).
7. Observe laser intersection point with the alignment microscope and move microscope to the intersection of laser and liquid jet (horizontal crosswire). Set the microscope's micrometer to zero at this position.
8. Move the microscope about 0.3 mm from the laser intersection point to the nozzle tip.
9. Move the nozzle down (Z axis control) until it hits the crosswire, i.e., is positioned 0.3 mm upstream from the laser intersection point.
10. Check the position of the forward scatter obscuration bar. It should block all of the laser beam.
11. Now the cytometer should be calibrated with standard particles. We use first glutaraldehyde-fixed CRBCs, because their forward scatter is very sensitive to misalignment, and doublets and triplets of CRBCs due to their low autofluorescence allow sensitive calibration of fluorescence. Use the following initial instrument settings for the CRBC standard: laser power: 150 mW (488 nm); trigger parameter: FSC; trigger threshold: channel 50.

Parameter	FSC	SSC	FL1	FL2
Photomultiplier Voltage	–	350	700	750
Linear gain	2	1	1	1

12. Run CRBC standard calibration, changing the following settings until the mean scatter and fluorescence reaches a maximum and the coefficient of variation a minimum: Y axis control of liquid stream, X axis control of stream, laser focus, focus of fluorescence microscope, position (height) of fluorescence microscope. Finally, you should obtain a FSC/FL1 dot plot such as that shown in Figure 5.

13. Calibrate compensation as described for the FACScan, using unstained and stained cells or using standard particles. StarCOMP software can be used for automatic calibration of sensitivity and compensation.

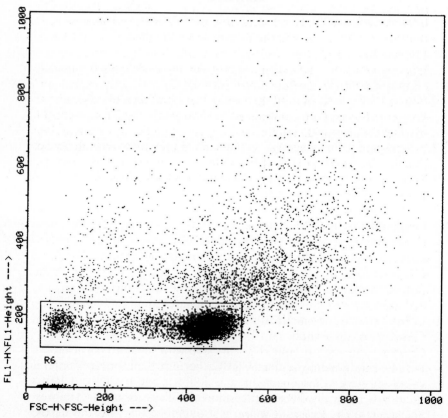

Figure 5. Forward scatter against green fluorescence (filter DF 530/30) of chicken standard

1. Remove sample tube, allow sheath backflush for at least 5 min. **Shutdown of**
2. Turn off laser power, but not laser cooling. **the flow-in-**
3. Run about 2 ml of a sample tube with 3–4 ml tissue-culture compatible **air cytometer** detergent (7X, Flow) with increased sample pressure (boost).
4. Turn off electronics.
5. Turn fluid main control to OFF position and turn off compressed air and vacuum.
6. Turn off laser cooling.

1.4 Tips, Tricks, and Troubleshooting

- Check that fluid control knob is in RUN position. **No cells**
- Check that pressure and vacuum is applied to the containers. **(FACScan)**
- Check for pinched tubings and air trapped in the sheath filter.
- The most common reason for this problem are clogged tubings or partially clogged flow chambers due to aggregates in the sample or previous shutdowns of the machine without flushing the tubing. The inner diameter of the sample tube is smaller than the inner diameter of the flow cell. Thus, usually the inlet tubing is clogged.
- Remove the sample tube and check whether the sheath flushes back. If not, wipe the end of the sample tube with a piece of research paper and check again. Switch the fluidics control to FLUSH for a few seconds and then to DRAIN. Wait until the flow cell is empty and switch to FILL. When the flow cell is filled again with sheath, switch to RUN and check for backflushing of sheath through the sample tube. If there is still no backflush, connect a syringe to the sample tube and try to drain out sheath. Backflushing should occur now; otherwise call the service. Before running samples again, flush the system with Chlorix.
- Remove aggregates from your sample if necessary.

- Check whether the PMTs are switched on. **No cells**
- Check threshold setting. Trigger signal should be switched to forward **(flow-in-air** scatter, and the threshold should be around channel 50. **systems)**
- Check whether the laser beam is hitting the liquid jet, and whether the reflection of the laser on the forward scatter obscuration bar is complete and indicates that the laser hits the middle of the liquid stream, i.e., the laser is reflected symmetrically to both sides.
- Check sample pressure.
- Check whether the air inlet of the sample tube is free.
- BOOST the sample pressure shortly and watch for forward scatter signals on the oscilloscope.
- If still no cells come through the nozzle, check for clogging of tubings and nozzle tip, as indicated above for the FACScan. Clean or exchange if necessary.

Optimizing forward scatter in flow-in-air cytometers
–   To obtain a good forward scatter, use an obscuration bar as small as possible. Use CRBCs for calibration of scatter.
–   A dot plot of forward scatter against fluorescence for a well-aligned FACStar is shown in the figure. Since forward scatter depends on the angle of measurement, other flow cytometers may give a rather different distribution [9].

Optimizing laserpower (flow-in-air cytometers)
The laser power for a specific dye can be optimized by running a sample with a mixture of unstained and stained cells and monitoring the mean of the unstained and stained cells for different power settings. Start at a laser power of 50 mW and end at 500 mW. The laser power is optimal for the greatest difference between stained and unstained cells in mean fluorescence (see Chap. 3).

Optical filters
–   Optical filters are available as longpass, shortpass, bandpass, and rejection band filters. Longpass filters allow light with long wavelength to pass and block light of short wavelength. Similar shortpass filters transmit light with short wavelength and block long-wavelength light. The filters are specified by the so-called cutoff wavelength. At this wavelength 50% of the light is transmitted and 50% is blocked.
–   Bandpass filters transmit a range of wavelengths. This spectrum is specified by the central wavelength and the bandwidth at the 50% of maximum transmission points (full-width half-maximum bandwith, FWHM band width).
–   Dicroic beam splitters are used for splitting light of different wavelengths. These are available as longpasses, shortpasses, or bandpass and used at a 45^0 angle of incidence.
–   Blocking of the strong laser light for measurement of fluorescence requires a minimum optical density of 5 at the laser wavelength. If excitation is near by the emission band, filters with very steep sides, ideally a rectangular-shaped transmission range of bandpass filters, are necessary. The DF series of bandpass filters from Omega (Omega Optical, Brattleboro, Vermont, USA) specified with bandwidth of 1.8 x FWHM bandwidth at the 0.001% (OD=5) transmission points and 3.2 x FWHM bandwidth at 0.0001% (OD=6) transmission points. The average transmission in the passband is more than 60%.

Sample filters
To prevent aggregates larger than the diameter of the nozzle tip from entering the sample line, sample filters should be placed at the beginning of the sample tube. Such filters can be purchased from the manufacturer or easily self-made from nylon gaze (30–60 μm mesh width) and yellow pipette tips. The first 5 mm of the tip is cut off and used to make a sample filter. The small cone is heated at the wide side on a piece of aluminium foil put on a hot plate until the plastic starts to melt. The cone is then pressed immediately onto a piece of nylon gaze which is placed on cold aluminium.

Do not press too much so that only the rim of the plastic is glued onto the nylon. Cut out the nylon with the tip on top. The sample filter has its wide side closed with nylon of defined pore size. Insert it with its narrow side into the end of the sample tube.

References

1. Melamed MR, Mullaney PF, Shapiro HM (1990) An historical review of the development of flow cytometry and sorters. In: Melamed M R, Lindmo T, Mendelsohn M L. (eds) Flow cytometry and sorting. New York: John Wiley & Sons, Inc. pp 1-9
2. Shapiro HM (1988) Practical flow cytometry. New York: Alan R. Liss, Inc.
3. Kachel V, Fellner-Feldegg H, Menke E (1990) Hydrodynamic properties of flow cytometry instruments. In: Melamed MR, Lindmo T, Mendelsohn ML (eds) Flow cytometry and sorting. New York: John Wiley & Sons, Inc. pp 27-44
4. Hamamatsu Photonics Deutschland GmbH D-8036 Herrsching; Catalog: photomultiplier tubes (December 1986)
5. Loken MR, Parks DR, Herzenberg LA (1977) Two color immunofluorescence using a fluorescence-activated cell sorter. J Histochem Cytochem 25:899-907.
6. Becton Dickinson (9/1988) User guide for the FACStar plus pulse processor. Becton Dickinson Immunocytometry Systems, 2375 Garcia Avenue, Mountain View, California 94043
7. Sharpless, TK and Melamed, MR (1976) Estimation of cell size from pulse shape in cytofluorometry. The Journal of Histochemistry and Cytochemistry 1976; 24(1): pp 257-264
8. Sharpless T, Traganos F, Darzynkiewicz Z, Melamed MR (1975) Flow cytofluorimetry: discrimination between single cells and cell aggregates by direct size measurement. Acta Cytologica 1975; 19(6): pp 577-581
9. Horan PK, Muirhead KA, Slezak SE (1990) Standards and controls in flow cytometry. In: Melamed M R, Lindmo T, Mendelsohn M L (eds) Flow cytometry and sorting. New York: John Wiley & Sons, Inc. pp 397-414
10. Becton Dickinson (3/1991) Setting up the FACScan for three-color flow cytometric analysis with PerCP conjugates, Becton Dickinson Immunocytometry Systems, Source Book Section 2.18.
11. Becton Dickinson (3/1987) FACScan AutoCOMP Software user's guide
12. Grandler W, Shapiro H (1989) Technical tutorial: logarithmic amplifiers, Cytometry Vol. 11, pp 447-450 (1990)
13. Schmid I, Schmid P, Giorgi JV; Conversion of logarithmic channel numbers into relative linear fluorescence intensity, Cytometry Vol. 9, pp 533-538 (1988)
14. Recktenwald D, et al. (1990) Biological pigments as fluorescent labels for cytometry. SPIE Vol.1206 (1990)

Part II Immunofluorescence

2 Conjugation of Fluorochromes, Haptens, and Phycobiliproteins to Antibodies

W. MÜLLER

2.1 Background

Cell-associated antigens can be recognized by antibodies. These bound antibodies can be visualized by several methods. No modification of the antibody is needed if one uses a second fluorochrome coupled antibody to develop the first uncoupled antibody bound to the cells. This indirect staining procedure has its limitation when one needs more than one color for the analysis. Biotinylation of the antibody is required if one wishes to use fluorochrome-coupled streptavidin reagents for the detection of the first antibody. Direct coupling of a fluorochrome to the antibody is required for the second or third color.

This chapter is divided into two parts; the first demonstrates the modification of antibodies with small fluorochromes or haptens, and the second shows how to cross-link phycobiliproteins to antibodies.

2.2 Coupling of Small Molecules such as Fluorochromes and Haptens to Antibodies

To couple small molecules to proteins such as antibodies, these small molecules must be used in an activated form so that they easily couple to proteins. For many small fluorchromes these molecules are available either as succinimide esters or as isothiocyanate. The main difference between the two groups of molecules in practical terms is the pH required for the optimal coupling reaction. Succinimide esters begin to react at a pH slightly higher then 7.0 while isothiocyanate requires a pH higher then 9.0. The use of succinimide ester is therefore preferable as the antibody does not has to incubated at high pH.

Small molecules such as biotin or fluorescein are chemically more resistant compared to phycobiliproteins. This allows, for example, the use of strong fixation procedures for the cells after staining. As activated small fluorochromes and haptens are commercially available, the procedure needed for antibody modification is very simple.

The quantum yield of a small fluorochrome such as fluorescein, however, is low compared to phycobiliproteins. This limitation can be overcome by using self-made biotinylated antibodies developed by a commerically available streptavidin coupled to phycobiliproteins. When using succinimide ester for an indirect staining method, one should try to obtain an ester which includes a spacer arm such as capronic acid. This spacer arm adds more flexibility to the small molecule and increases the chance that groups added to the antibody are acceptable by the second-step reagent.

2.2.1 Materials

Succinimide esters: There are several forms of succinimide esters, such as fluorescein FLUOS (Boehringer, Mannheim), biotin (e.g., NHS-LC biotin; Pierce, Molecular Probes), and haptens like digoxigenin (Boehringer, Mannheim).

Isothiocya-nates Several isothiocyanate-coupled fluorochromes are available from many companies: fluorescein isothiocyanate (Calbiochem), tetramethyl-rhoda-mineisothiocyanate (Molecular Probes), and sulforhodamine 101-iso-thiocyanate (Texas red, Molecular Probes).

Buffers
- Coupling buffer for succinimide esters: (prepared fresh) 0.1 M sodium hydrogen carbonate pH 8.0, antibody (1 mg/ml or higher)
- Coupling buffers for isothiocyanates: isothiocyanate requires high pH for the coupling reaction (pH 9.4). This high pH could be harmful for the antibody. Two step dialysis of the antibody prior to coupling increasing the pH for only a short time reduces the problem to an acceptable level.
- Buffer A: 0.1 M boric acid, 0.025 M sodium tetraborate, 0.075 M sodium chloride, adjust pH to 8.4 using sodium hydroxide.
- Buffer B: as buffer A but adjust the pH to 9.5.

Miscellane-ous
- Dimethylformamide (DMF)
- Dialysis tubes
- Reagent glass tubes (approximately 10 ml)
- Small stirring bar
- Magnetic stirrer

2.2.2 Method

All reactions are performed in reagent glass tubes at room temperature.

1. The antibody is adjusted to a protein concentration of 1 mg/ml in coupling buffer and dialyzed extensively against coupling buffer – when using isothiocyanate-activated molecules, dialyzed extensively against coupling buffer A. One hour before the coupling the antibody in the dialysis bag is transferred to coupling buffer B. This step removes substances such as sodium azide, which could be present in the antibody

preparation and would inhibit the reactions, and it adjusts the correct pH for the coupling reaction.

2. The dialyzed antibody is placed in a reagent glass tube, and a small magnetic stirring bar is added.

3. The activated fluorochromes or biotins are dissolved in dimethylformamide: 1 mg/ml for succinimide esters, 2 mg/ml for fluorescein isothiocyanate, 3 mg/ml for tetramethylrhodamine isothiocyanate, and 10 mg/ml for Texas red.

4. Add 100 µl of this solution per 1 ml antibody solution (1 mg/ml) and stir the reaction for 1 h at room temperature.

5. Separate uncoupled from coupled small molecules: Either dialyze the reaction mixture overnight against phosphate-buffered saline (PBS)/azide or purify the coupled antibody over a PD-10 column (Pharmacia; desalting column). For the latter equilibrate the PD-10 column with PBS/azide (20 ml/column). Apply the reaction mixture to the column (maximum 2 ml/column) and elute the labeled antibody with PBS/azide. The antibody comes first!

6. Determine the coupling ratio: The conjugation rate of fluorochrome conjugates can be easily determined in a photometer. The formula given below gives approximate values only, mainly because the protein concentration is not measured very exactly. The values obtained, however, are very useful for contolling the coupling reaction.

- **Fluorescein:**
  Determine OD 280 and OD 495
  molar ratio of fluorescein to protein   protein concentration (for Ig):

$$F/P = \frac{2.87 \times OD\ 495}{OD\ 280 - (0.35 \times OD\ 495)} \qquad mg/ml\ Ig = \frac{OD\ 280 - (0.35 \times OD\ 495)}{1.4}$$

- **Texas red (sulforhodamine 101)**
  Determine OD 280 and OD 595
  molar ratio of Texas red to protein   protein concentration (for Ig):

$$TR/P = \frac{2.28 \times OD\ 595}{OD\ 280 - (0.5 \times OD\ 595)} \qquad mg/ml\ Ig = \frac{OD\ 280 - (0.5 \times OD\ 595)}{1.4}$$

- **Tetramethylrhodamine**
  Determine OD 280, OD 515, and OD
  595 molar ratio of tetramethylrhodamine to protein   protein concentration (for Ig):

$$R/P = \frac{9.25 \times OD\ 555}{OD\ 280 - (0.56 \times OD\ 515)} \qquad mg/ml\ Ig = \frac{OD\ 280 - (0.56 \times OD\ 515)}{1.4}$$

7. Titrate out the modified antibody on test cells

8. Freeze in convenient small aliquots at -70°C

2.2.3 Tips, Tricks, and Troubleshooting

Antibody precipitates – Some of the antibodies might precipitate during the reaction. When this happens, remove these aggregates by centrifugation in an Eppendorf centrifuge before the final dialysis step.

Loss of specific binding – The antibody might loose specific binding. This can be tested by staining cells with the antibody before and after the coupling and developing this staining by an antiserum directed against the antibody (indirect staining). If this happens, try to use smaller amounts of the coupling reagent.

No coupling – The antibody is not coupled. This could be due to insufficient dialysis of the antibody prior to coupling (sodium azide, Tris). The activated molecules may be broken down (keep dry at -20°C), degraded dimethylformamide, wrong pH.

Non specific binding – The antibody binds nonspecifically. Make sure that the antibody preparation is of good quality (purified antibody without other protein contaminants such as transferrin). Try to use smaller amounts of the coupling reagent. Change from a succinimide ester to isothiocyanate or vica versa. If you wish to sterilize the antibody conjugate by filtration, use low protein binding membranes; otherwise you will loose most of your conjugates.

Separate antibodies according to coupling ratios – The staining property of an antibody conjugate can be improved by separating the conjugate over an ion exchange column:

Material 2 ml syringe glass wool DE 52 (Whatman)

Binding buffer: 10 mM phosphate, 1 mM NaN$_3$, pH 7.4

Elution buffers: 10 mM phosphate, 5,10,20, and 50 mM NaCl, 1 mM NaN$_3$, pH 7.4

Method Equilibrate DE 52 with binding buffer. Put glass wool into the bottom of a 2 ml syringe. Add 1 ml of the preequilibrated DE 52. Equilibrate the fluorochrome antibody complex with binding buffer (10 mM phosphate, 1 mM NaN$_3$, pH 7.4.) using either the PD-10 column or dialysis. Apply the complexes to the DE 52 column. Elute the bound proteins with several steps: 5,10,20,50 mM NaCl in 10 mM phosphate, 1 mM NaN$_3$, pH 7.4. Noncoupled antibody elutes first, complexes later, huge complexes last. For each fraction determine the F/P ratio in a photometer. Test the various fractions by staining.

– Sometimes it is possible to remove antibodies which lost binding activity   Repurification
due to the coupling procedure to the antigen from intact antibodies by   of the anti-
affinity chromatography. For this the antigen is coupled to Sepharose.   body by affinity chro-
The antibody is added to the column. Nonbinding antibody runs   matography
through. Bound antibody can be eluted by low pH. This does not work
for phycobiliprotein coupled antibodies.

2.3 Cross-linking of Phycobiliproteins to Antibody

The following procedure is a very simple method for cross-linking phycobiliproteins to antibodies or to streptavidin [3]. To avoid complicated
purification of the complexes formed by cross-linking later, the strategy of
the coupling method is to use the phycobiliprotein in a two- to fourfold
molar excess over the antibody. By this all antibody molecules become
coupled. The uncoupled phycobiliproteinis are removed after staining of
the cells. Phycobiliproteins are very sensitive to light! Try to cover all tubes
and columns to protect the protein from light. Especially light from neon
lamps is very destructive.

2.3.1 Material

– PD 10 columns (Pharmacia)
– Reagent glass tubes
– Coupling buffer 100 mM sodium phosphate, 50 mM sodium chloride
  pH 6.8
– Antibody (4 mg/ml or higher)
– 2-Iminothiolane (Sigma I-6256)
– GMBS = N-c-maleimiddobutyryloxy succinimide (GMBS; Calbiochem
  442630) MW 280.2
– Phycobiliproteins (e.g, R-phycoerythrin; Calbiochem, Boehringer; or
  allophycocyanin, Calbiochem).

2.3.2 Method

The amounts given are for coupling of about 1.5 mg IgG

– Exchange 4 mg phycoerythrin or 3 mg allophycocyanin to coupling   Thiolation of
buffer using PD-10 column (preequilibrated with coupling buffer).   phycobilipro-
– Put the exchanged phycobiliprotein into a reagent glass tube. Prepare   tein
a solution of 1 mg 2-iminothiolane/0.1 ml coupling buffer.

- Add 30 μl of this solution to the phycobiliprotein solution.
- Stir the reaction mixture for 2 h at room temperature.
- Isolate thiolated phycobiliproteins using PD-10 column (preequilibrated with coupling buffer).

Labeling of antibodies with GMBS
- The antibody solution is adjusted at a concentration of 1.5 mg/0.45 ml to coupling buffer (either by dialysis or, again, using the PD-10 column.
- The liquid eluting from the column is collected in 0.5 ml aliquots. The presence of protein in each fraction can be easily determined by testing small aliquots (10 μl sample in 200 μl Bradford's reagent) in the Bradford assay [2] on a microtiter plate. Dissolve 1 mg GMBS in 180 μl DMF; use 4.5 μl/0.45 ml antibody (1.5 mg antibody). Stir the reaction mixture for 1–2 h at room temperature; remove excess GMBS with PD-10 column.

Forming cross-linked phycobiliprotein-antibody complex
Add slowly (over a 30 min period) 1 M equivalent of GMBS-labeled antibody to 2–4 M equivalents of 2-iminothiolane modified phycobiliprotein. Example: for 1.5 mg antibody use 4 mg phycoerythrin or 3 mg allophycocyanin to allow for a 2x excess of phycobiliprotein. (The molecular weight of an antibody of the IgG class is about 140 kDa, of phycoerythrin about 240 kDa, and allophycocyanin about 120 kDa).

Slow addition of one solution to another is technically very easy. Fix a syringe with a small needle (20 g) above the reaction glass tube. Add the solution that you want to add to the syringe. Press slightly onto the syringe upper opening (the flow must be started by this little push) using the syringe plunger. The solution should now drop slowly into the reagent glass tube. To mix the incoming solution with the solution present in the reagent glass tube place a small magnetic stirring bar into the tube and place it above a magnetic stirrer.

Continue to stir the reaction mixture for 1–2 h at room temperature.

Testing Test the modified antibody on cells. Usually there is no need to further purifiy the mixture. We wash the cells twice after the staining to remove uncoupled phycobiliproteins.

Freezing Freeze small aliquots at -70°C. Test a small aliquot first as not all protein conjugates survive freezing and thawing. In case of problems keep aliquots at +4°C in the dark.

2.3.3 Tips, Tricks and Troubleshooting

Purification of phycobiliprotein-anti-body complexes
- If you want to sterilize the antibody-conjugate by filtration, use low protein binding membranes; otherwise you will loose most of your conjugates. If there is a problem by nonspecific background staining or weak staining, it might help to purify the phycobiliprotein-antibody complexes.

2 ml syringe glass wool Sephadex G10 hydroxyapatite

Binding buffer: 1 mM phosphate, 100 mM NaCl, 1 mM NaN$_3$, pH 7.0

Elution buffers: 10,40,80, and 100 mM phosphate, 100 mM NaCl, 1 mM NaN$_3$, pH 7.0

Mix 1 part hydroxyapatite with 1 part Sephadex G10.
Put glass wool into the bottom of a 2 ml syringe.
Add 1 ml hydroxyapatite/Sephadex G10 mixture.
Equilibrate this column with binding buffer (1 mM phosphate, 100 mM NaCl, 1 mM NaN$_3$, pH 7.0.)
Equilibrate the phycobiliprotein-antibody complex with binding buffer (1 mM phosphate, 100 mM NaCl, 1mM NaN$_3$, pH 7.0.) using either the PD-10 column or dialysis.
Apply the complexes to the hydroxyapatite column.
Elute the bound proteins with several steps: 1, 5, 10, 40, 80 mM phosphate in 100 mM NaCl, 1 mM NaN$_3$, pH 7.0.
Free phycoerythrin elutes first, complexes later, largest complexes last.
Test the various fractions by staining (antibody-phycoerythrin complex should elute at 40 mM phosphate).

References

1. Kitagawa T, et al. (1983) J. Biochem. 94, 1160.
2. Bradford MM (1979), Anal. Biochem. 72, 248.
3. Hardy RR (1986) Handbook of experimental immunology, Volume 1, chapter 31, ed. D.M. Weir. 1986.

3 Immunofluorescence: Basic Considerations

A. Radbruch

3.1 Background

Immunofluorescence involves the staining of cells with antibodies and other specific ligands directly or indirectly labeled with fluorescent dyes. This offers fascinating possibilities for the characterization of cells according to the expression of protein or polysaccharide markers specific for differentiation or activation stages. A few basic considerations will help one to understand the important parameters and to optimize the specificity and sensitivity of immunofluorescence staining. Of the three steps in immunofluorescence analysis–staining, measurement, and data evaluation–the first, preparation of cells and staining, is probably most important and therefore receives the most attention.

3.2 Staining

3.2.1 Preparation of Cells

As a matter of principle, the manipulation of cells should be kept to a minimum because of the risk of introducing artifacts. However, since flow cytometry requires single-cell suspensions, and cells are analyzed one by one, it is obvious that attached or connected cells must be individualized either mechanically or enzymatically. Care must be taken that the enzymes used for separating the cells do not attack the antigens to be analyzed later. Other manipulations frequently used are preenrichment of rare cells and fixation of stained cells prior to analysis to avoid phenotypic changes and to minimize the risk of infection.

Natural single cell suspensions, such as blood or bone marrow, can be stained without any further manipulation of the sample. Most present protocols, however, include preenrichment steps, for example, the depletion of erythrocytes for analysis of leukocytes from blood. This saves reagents due to the smaller staining volume and improves discrimination of rare cells and may also remove dead cells and other cells that could be stained unspecifically. Preenrichment procedures such as density gradient centrifugation, lysis, and magnetic cell sorting are discussed in detail in Chaps. 14 and 15. In choosing a method for preenrichment, however, one should keep in mind that this is a manipulation which may also affect the cells to be analyzed, especially if the cells are from pathological rather than normal conditions.

3.2.2 Direct Versus Indirect Staining

Direct staining with fluorochrome-conjugated antibodies against specific cellular determinants is preferable, if possible, at least for routine use. Manipulation of cells is minimized, the staining is easy to control, and it allows the combination of several markers for multiparameter analysis, mixing the various staining reagents in one staining step. Conjugation of even small amounts of purified antibodies with fluorochromes is fast and easy to perform in any laboratory (see Chap. 2).

For „negative" control–usually antibodies of the same class, conjugated to the same fluorochrome as the specific antibody but binding to an irrelevant antigen–are used to stain another aliquot of the cells (isotype control). Although „state of the art", this is not a very good control because it is dependent on the quality of the conjugation. A „cellular control" is preferable, using the specific antibody to stain cells not expressing the antigen. In heterogeneous cell populations, such as blood cells, negative cells are usually present and can be compared to unstained cells to evaluate background staining. Cells expressing the antigen in question should be used as „positive" control.

Indirect staining with unconjugated antibodies against specific cellular determinants, detecting the antibodies with second, fluorochrome-conjugated, generally polyclonal anti-antibodies, must be used if the first antibody cannot be purified. Control of indirect staining is not trivial, and multiparameter analysis is even less so. Combination of several parameters requires highly purified isotype-specific second antibodies or complete blocking of the free binding sites of the second antibody before staining of the next parameter, i.e., increasing the number of staining steps from one to four or more. Every reagent used for staining or blocking must be controlled independently in the way indicated above, and in case of cross-reactions of the polyclonal second antibodies may require extensive absorptions on cells or immobilized proteins.

An advantage of indirect over indirect staining is that it can be used to improve the sensitivity of detection since several second antibodies may bind to one first antibody, thus increasing the number of fluorochromes per specific antigenic determinant. A further increase in sensitivity can be obtained if the first antibody is conjugated to the same fluorochrome as the second. High-specificity and sensitive, easy-to-control indirect staining is obtained with first antibodies conjugated to haptens such as nitro-iodophenylacetyl, digoxigenin, or biotin, detected with fluorochrome-conjugated hapten-specific second antibodies or avidin. The biotin/streptavidin system has the advantage of readily commercially available streptavidin conjugates, not only for immunofluorescence but also for histochemistry or serology. Also, it is easy to conjugate even small amounts of purified antibodies (>100 µg) quantitatively to biotin (Chapter 2).

3.2.3 Staining Parameters: General Considerations

Standardization of concentrations of cells and antibodies is required to obtain reproducible results. Concentrations should be low to support specific high-affinity antigen-antibody interactions over low-affinity cross-reactions but high enough to obtain staining as bright as possible. Manipulation of cells should be kept to a minimum, i.e., the number of staining and washing steps should be as low as possible to avoid loss and damage of cells.

3.2.4 Concentration of Cells

Cell number and concentration should always be chosen to allow staining at the optimal concentration of cells and antibodies, determined by titration beforehand. The lower limit is approximately 10^5 cells in 10 µl since about 10^5 cells are required for flow cytometric analysis, recording 10^4 cells. The lower limit of volume for staining is 10 µl since after washing of the cells a few microliters of washing fluid always remains in the tube, changing the concentration of the staining antibody in an unpredictable way. We use staining volumes not of 10 but of 20–100 µl with 0.2–1x10^7 cells, i.e., cell concentrations of 10^8 per milliliter. For staining of larger numbers of cells the volume and not the concentration of cells or stain should be increased. In principle, all cell concentrations refer to the number of positive cells, and changes in volume due to additional negative, nonstainable cells can be neglected. For the staining of rare cells, however, the volume of negative cells must be taken into consideration in determining the optimal concentration of cells and stain.

3.2.5 Concentration of Staining Reagents

Staining reagents should always be titrated. Titration helps to save reagents and to optimize the staining itself. Concentrations that are too low will result in poor discrimination between positive and negative cell populations, i.e., overlapping populations (see below), making statistical evaluation difficult or impossible. Concentrations that are too high lead to unspecific staining of negative cells or subpopulations on the basis of low affinity cross-reactions.

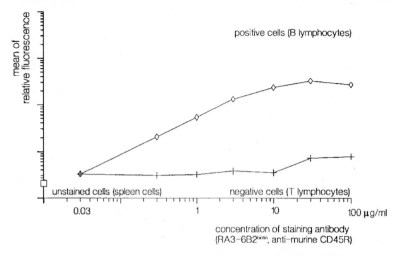

Figure 1. Titration of staining reagents for flow cytometry. RA3-6B2[biotin] (anti-murine CD45R) was titrated on murine spleen cells, staining 2×10^6 cells in 20 μl staining reagent at concentrations as indicated for 10 min, washing once, then staining with streptavidin-fluorescein at 3 μg/μl, 20 μl for 2×10^6 cells for 10 min on ice, washing once, and analyzing in a FACScan flow cytometer. Murine B lymphocytes, about 50% of the lymphocytes in spleen, are positive for CD45R; T lymphocytes, the other 50% of lymphocytes, are not. For statistical evaluation, gates were set between positive and negative populations, and the mean of relative green fluorescence was determined for the gated populations by FACScan research software (Becton-Dickinson) and plotted versus the concentration of staining reagent. At 0.1 μg/ml positive and negative populations were overlapping and thus not evaluated. The optimal concentration of RA3-6B2[biotin] is 10 μg/ml for 10^6 cells (50% of 2×10^6 cells, i.e. positive B lymphocytes) stained in 20 μl. For staining of higher cell numbers, the volume must be scaled up linearly, not the concentration. (Data from G. von Hesberg and A. Radbruch, Cologne)

Titration can be performed by flow cytometry (Fig. 1), staining a mixture of positive and negative cells under standardized conditions at given concentration of cells, volume, time, and number of washing steps, with varying concentrations of the staining reagent to be titrated. The mean fluorescence of positive and negative populations is then plotted against the concentration of the staining reagent. The optimal concentration of the staining reagent is usually the concentration at beginning of the plateau of saturation. It is sometimes useful to employ lower concentrations, for example, in case of problems with cross-reactions, or higher concentrations, such as in case of low antigen density.

3.2.6 Staining Time

High-affinity antigen-antibody reactions occur fast, while longer staining times favor low-affinity cross-reactions or adsorptions. Thus, staining times should be as short as possible. Most staining reagents reach about

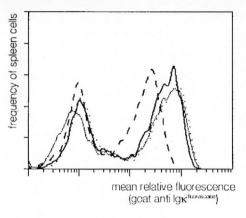

Figure 2. Determination of staining time. Murine spleen cells were stained with fluorescein-conjugated goat anti-murine immunoglobulin κ light chains, labeling most of the B lymphocytes, i.e., about 50% of the spleen cells. At defined time intervals, aliquots were removed, diluted (see following figure), and immediately analyzed at log amplification in a FACScan flow cytometer. The smoothed fluorescence histograms (number of cells versus fluorescence intensity) of various time points are overlaid to illustrate the differences in staining intensity. After only 1 min positive and negative populations are separated. Maximum staining intensity is reached at about 5 min. Thus, for polyclonal antibodies a staining time of 10 min is generally sufficient. Longer staining times do not improve the staining intensity but support low-affinity cross-reactions. For monoclonal antibodies, in the case of low affinity, the optimal staining time may have to be prolonged, as should be determined in a test staining such as that shown here. For cytoplasmic staining with antibodies or other proteins, the staining (and washing) times must be much longer, 1 h or more, depending on the fixation and the size of the cells, since the staining reagents must penetrate the fixed cytoplasm (see Chap. 5). – –, 1 min; ——, 5 min; - - -, 15 min; · · · ·, 30 min

90% maximum logarithmic staining after 1–5 min (Fig. 2). Doubling the time for safety reasons gives a standard staining time of 10 min. Staining of cytoplasmic parameters of fixed cells (Chap. 5) requires longer times because the staining reagents must diffuse through the cytoplasm. Staining and washing times of about 1 h are adequate for intracellular immunofluorescence. Staining times must also be prolonged if antibodies are not used but rather antibodies conjugated to large particles, such as liposomes or Dynabeads.

3.2.7 Staining Temperature

Staining of live cells at 4°–12°C („on ice") minimizes physiological influences. At room temperature or 37°C most cells are able to modulate expression of the antigenic receptors, especially if these are cross-linked by (staining) antibody. Lymphocytes can „cap" their receptors with the staining antibodies, throw off the cap into the medium, and appear „negative" then. Capping can be inhibited by sodium azide (0.03% in the staining medium), which reversibly blocks cell physiology.

Figure 3. Minimizing washing steps. Murine spleen cells were stained with fluorescein-conjugated goat anti murine immunoglobulin κ light chains, as described for the preceding figure. Cells were either not washed at all, diluted only 1:50 for analysis, or washed once, i.e., diluted 1:50, spun down, the supernatant removed completely, and the cell pellet resuspended in fresh buffer. The cells were then analyzed at log amplification in a FACScan flow cytometer. The smoothed fluorescence histograms (number of cells versus fluorescence inten-

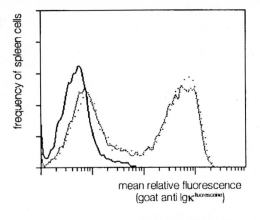

sity) are overlaid to illustrate the differences in staining intensity. Since diluted and washed samples show similar fluorescence distributions, washing is actually not required in this particular system. In most staining systems, washing once is sufficient. Further washing steps do not improve the quality of the staining but result in loss of cells. As many as 10% of cells can become lost per washing step.

——, Cells not stained; - - -, stained cells, diluted; · · · ·, stained cells, washed once

3.2.8 Washing

Like all other manipulations, washing should be restricted to the minimum. Washing–removal of unbound antibodies by centrifugation of the cells, removal of the supernatant, and resuspension of the cells in fresh medium– always leads to a loss in cell numbers, frequently those of selective subpopulations of small or less dense cells. Even under optimal conditions, 10% of cells can be lost per washing step. When the staining is performed with titrated, i.e. low concentrations of antibodies, one washing step is usually sufficient (Fig. 3). For indirect staining, washing twice efficiently prevents reaction of the first reagent with the second one in solution, which would result in suboptimal staining.

3.2.9 Fixation

Like all other manipulations, fixation may have unpleasant side effects. One such side effect is that it becomes difficult to distinguish cells that were dead before staining and therefore stain unspecifically. Such cells are easy to display in live material by propidium iodide staining. After fixation they can be identified only according to scatter and staining with LDS 751, which is not very easy (Chap. 4). For analysis of rare cells, the distinction between rare positive cells and unspecifically stained dead cells is crucial. Such samples are best analyzed unfixed, in the presence of propidium iodide, gating out dead cells according to scatter and propidium iodide fluorescence (Chap. 17).

Nevertheless, for routine analysis of subpopulations cells are usually fixed to limit the risk of infection and to standardize conditions of analysis. For staining of cytoplasmic and nucleic antigens, fixation is obligatory because otherwise the staining antibodies cannot penetrate the cell membrane.

Standard fixation for immunofluorescence is fixation with 0.5%–2% formaldehyde. Since this fixation does not preserve nucleic acids very well, for analysis of RNA or DNA (e.g., of sorted cells) fixation with 70% methanol or ethanol/acetic acid (95/5) should be preferred. For staining of cytoplasmic or nucleic antigens in formaldehyde fixed cells, permeabilization of the membranes is required, for example by 0.1% NP40 or 0.5% saponin (Chap. 5).

3.3 Measurements

From its beginning, flow cytometry has required a certain technological effort; however, the days of self-made machines are now a thing of the past. Today, several manufacturers offer state-of-the-art equipment at prices comparable to those of good microscopes. Analysis of forward and orthogonal scatter and three fluorescence light parameters per cell is now standard. At first sight, five parameters may seem too many for most applications; however, every additional parameter opens new possibilities, for example, increasing the resolution of „gating," i.e., the preselection of cells for analysis according to staining with vital dyes or immuno-fluorescence. This is extremely important for the analysis of rare cells, enabling, for instance, the direct analysis of leukocytes stained in blood without any further manipulation if the leukocytes are preselected elec-tronically (gated) not only by scatter but also by immunofluorescence (e.g., according to CD45 staining).

Most modern flow cytometers are stably aligned, which had been a problem in the early days and remains so for the flow-in-air cell sorters. They are also easy to operate. Calibration (optimization of sensitivity) and compensation (correction of fluorescence overspill; Chap. 1) are per-formed semiautomatically with the help of standard particles. Measure-ment and data storage constitute no problem. Preselection of cells during measurement (live gating) is a dangerous option and should be used only with caution. Live gating always results in loss of information because not all cells of the sample are analyzed. In basic research and clinical routine this may lead to misinterpretation of unexpected and pathological situa-tions.

While the intensity of the light scatter of cells is analyzed on the basis of linear amplification, the intensity of immunofluorescence is analyzed upon logaithmic amplification (Fig. 4). The reason is that logarithmic amplifica-tion over four decades which is state-of-the-art technology, allows one to use a single gain, i.e., the same machine settings for up to 10 000-fold differences in fluorescence, which is about that which can be observed with present-day staining technology. Thus, relative fluorescence of various

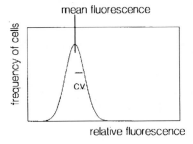

Figure 4. Mean fluorescence and CV. In flow cytometric analysis of immunofluorescence, the cells (usually 10 000) are classified according to relative, logarithmically amplified fluorescence intensity in 256, 512, or 1024 channels, depending on the setup of the machine. Due to variation in the cells and in the measurement a population of homogeneously stained cells is distributed symmetrically around a mean value of relative fluorescence (arithmetic mean for linear amplification, geometric mean for logarithmic amplification), with a dispersion described as coefficient of variation (CV, standard deviation divided by the mean and multiplied x 100). (Histogram produced by computer simulation by K.L. Meyer)

cells can be compared directly. Also, the variation in staining upon logarithmic amplification resembles that of a gaussian normal distribution, which makes it easier to analyze the fluorescence distributions (see below). Finally, logarithmic amplification emphazises the biologically usually more important differences between cells expressing little antigen, while differences between cells expressing a great amount are neglected.

3.4 Data Evaluation

3.4.1 Gating

In flow cytometric analysis all available optical parameters of a particle are recorded if the forward scatter light of this particle exceeds a preset threshold value (trigger). In principle, every other parameter can be used as well to trigger the analysis. For data evaluation, further constraints can be put on the particles to be analyzed, setting lower and upper thresholds for other parameters. This electronic preselection (gating) is used to exclude unspecifically stained particles from analysis (e.g., dead cells or debris) or to enrich rare cells electronically to obtain better statistics.

Gating can be risky if the cells behave abnormally, for example, in pathological situations. Thus, gating requires substantial expertise and can result in unfortunate mistakes. Since gating parameters and gates are frequently not documented, these mistakes are hard to identify. In any case, establishing gates should be controlled by staining of independent parameters, a further advantage of the additional parameters available on modern machines. This control should be obligatory for the definition of „standard" gates, such as the lymphocyte scatter gate, which can easily be controlled by CD45 staining as all CD45[+] cells and only these should fall into the scatter gate. If no „independent" parameters are available, gates could be verified by cell sorting, sorting out gated cells and analyzing them by microscope or any other suitable method.

3.4.2 Plotting of Data

From the multitude of data plots available, only a few are relevant for immunofluorescence (Fig. 4). One-dimensional histograms are best suited for illustrating the logarithmically amplified intensity of staining (mean fluorescence and coefficient of variation, CV), i.e., for emphazising the quantitative aspects of flow cytometry. Two-dimensional plots (Fig. 5),

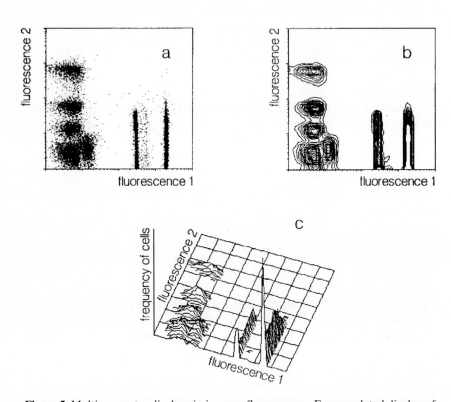

Figure 5. Multiparameter displays in immunofluorescence. For correlated display of two immunofluorescence parameters either dot plot (a) or contour plot (b) are most commonly used. In the dot plot (a) every cell is displayed as a dot according to its classification for intensity of fluorescence 1 and 2 in the coordinates of fluorescence 1 (x-axis) and 2 (y-axis). Cells with the same intensities are superimposed. This makes the dot plot the display of choice for analysis of rare cells, while large cell populations are not well resolved. Contour plots are derived from dot plots by delineating areas of equal density, either linearly, as shown here (1% linear) or logarithmically, or according to probability ([1] Chap. 30). Contour plots are the display of choice for analysis of populations of frequent cells, giving high resolution even if the populations are not well separated by staining. The three-dimensional histogram (c) tends to confuse more than to clarify. It is useless for illustration. The data shown here were obtained from analysis of a mixture of beads from Flow Cytometry Standards Corporation on a FACStar+, evaluated with Lysis II software. Seven beads were used, containing no fluorochrome, 1.1×10^4 fluorescein equivalents (fe), 5.1×10^4 fe, 4.6×10^5 fe, 5.6×10^4 phycoerythrin equivalents (pe), 10^5 pe and 10^6 pe. Such mixtures of beads should be routinely used to check and align sensitivity and compensation of flow cytometers (see Chap. 1)

either as dot or as contour plots, give a better impression of the interrelated expression of two parameters per cell. In a dot plot, every cell is plotted as a dot according to the intensity of its staining between the coordinates of the two parameters. Rare cells are emphasized and frequent cells are not, because areas with many cells appear merely black, and subpopulations are not resolved. Delineating areas that contain classified frequencies or numbers of cells between the coordinates of the two parameters results in contour plots, which tend to give a detailed and graphic picture of densely populated areas while rare cells can become lost. Three-dimensional plots, like three-dimensional histograms or three-parameter dot plots (clouds) are seldom used in immunofluorescence, as they may overtax our visual imagination, at least in two-dimensional display such as on paper or screen.

Sometimes the data are no longer shown, and only the frequency of subpopulations is given in numbers and tables. This can be dangerous, especially in the case of overlapping or shifted populations, as discussed below.

3.4.3 Statistical Evaluation

Statistical evaluation of immunofluorescence analysis is not very demanding. Cell populations, as defined by scatter and fluorescence staining, are described by the relative mean and CV of a particular fluorescence parameter (Fig. 4) and the number of cells in that population. This evaluation is possible, however, only if (a) the distribution of fluorescence intensities for the population analyzed varies symmetrically around the mean value, and (b) the whole population is visible „on scale," i.e., not summed in the first (for negative cells) or last channel, which would make it impossible to determine values for the mean or CV of fluorescence or for the autofluorescence of negative cells. For determining the relative fluorescence of positive cells the relative fluorescence of autofluorescent negative cells should be the reference.

The mean is defined as the arithmetic mean for linear scale and geometric mean for logarithmic scale. The CV (in percent) can be calculated independently of the mean fluorescence, as half the width of the distribution at 0.6 times maximum height:

$$CV = 100 \times (10^{(a-b)/2\,S}-1) = 100 \times (10^{2\,SD/2\,S}-1), SD = \log\left[(CV/100 + 1) \times S\right]$$

with S = channels per decade (usually 256), a = upper channel with 0.6 maximum height, and b = lower channel with 0.6 maximum height [4]. The variation observed and described as CV is the result of both biological variation (e.g., size of cells, density of target antigen, and staining antibody) and variation of measurement (e.g., focusing, orientation), and should describe a logarithmically symmetrical distribution around the mean of relative fluorescence (Fig. 4).

Determination of the absolute number of antigen molecules per cell is difficult today and requires a reference method for calibration of the relative immunofluorescence. The frequency of positive or negative cells

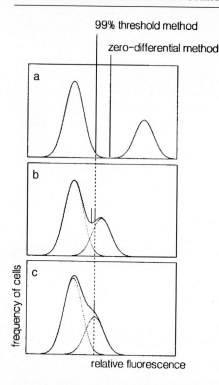

99% threshold method

zero-differential method

a

b

frequency of cells

c

relative fluorescence

Figure 6. Evaluation of separate and overlapping populations. A mixture of cells, the normal sample for immunofluorescence, is easy to analyze, if the populations of positive and negative cells on the histogram are separate (a; see also figures on determination of staining time and on minimizing washing steps). An analysis gate, either including all cells „not negative" (99% threshold method) or set at middle distance between positive and negative populations, defines areas of the fluorescence scale, for which the mean of fluorescence and CV are determined (see figure on mean fluorescence and CV). Thus the frequency of positive and negative cells and their respective mean fluorescence and CV describe the sample with respect to the parameter of immunofluorescence. Staining should be optimized to achieve this clear and easy-to-analyze situation. (Histogram produced by computer simulation by K.L. Meyer) Bivariate or asymmetrical histograms (b and c) are not that easy to analyse. such histograms could represent kinetic transitions (e.g. activation markers) or a defined number of distinct populations, overlapping in immunofluorescence due to weak expression of antigen or suboptimal staining. The method of choice for evaluation of such histograms would be curve fitting. Curve fitting programs are available from Partec or Verity Software. For daily routine, the „zero- differential"-method can be used, as long as the histograms are bivariate and contain about equal numbers of positive and negative cells. Setting the analysis gate between „positive" and „negative" parts of the histogram, at the point of „zero increment", will allow an estimation of the frequencies of positive and negative cells, because „false positive" and „false negative" cells will level out, as long as the populations occur in about equal frequency. Bivariate histograms with one population being rare and asymmetrical histograms (c) have to be evaluated by curve fitting. In terms of experimental strategy, however, additional parameters, optimizing the staining and, in case of rare cells, preenrichment (see Chap. 14) could improve the resolution of the analysis, allowing evaluation without or controlling curve fitting

can be turned in absolute numbers only if the number of cells is known from which the sample has been taken.

Under optimal conditions, i.e., with perfectly separated populations (Fig. 6), the statistical values for mean and CV of relative fluorescence and the frequency of cells of the various populations are easy to obtain. The distinct populations are defined by analysis gates (lower and upper thresholds of intensity of fluorescence), and the statistical values are determined for the cells within the windows. Frequently the threshold is set to include 99% of the cells of the negative control, with 1% showing higher fluorescence. With separate populations, and even less so with nonseparate populations, there

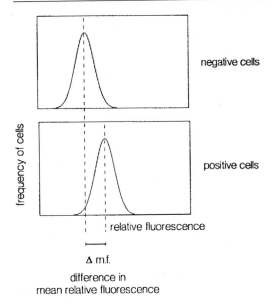

Figure 7. Evaluation of shifted populations. All the cells of a population that stain homogeneously, resulting in a symmetrical fluorescence distribution, defined by mean of fluorescence and CV, are positive, including those classified as dim as the negative control. This classification is due to lack of resolution of the flow cytometric analysis and variation in staining and measurement. The increase in fluorescence staining relative to the negative control is described by the difference in mean fluorescence and CV between negative control and positive, shifted population. The 99% threshold method of data evaluation (see preceding figure) cannot be applied to shifted populations. (Histogram produced by computer simulation by K.L. Meyer)

is no reason to use this method. Setting the threshold anywhere between the populations always gives more accurate frequencies.

It is not very easy to evaluate overlapping and shifted populations (Figs 6 and 7). Frequently, also for these nonseparate populations, threshold gates are used for evaluation, setting the thresholds to include 99% of the „negative" control, a sample of cells stained with an irrelevant antibody, and considering only those cells with higher fluorescence as „positive." For nonseparate populations, this procedure simply gives wrong statistics. This frequent mistake is one of the reasons that it should be made clear in publications, that statistical immunofluorescence data, when presented in tables, refer to histograms with separate populations.

How does one evaluate nonseparate populations? Of the nonseparate fluorescence histograms, shifted populations (Fig. 7) are easy to evaluate because they differ from the negative control only by the mean fluorescence, and the difference in mean fluorescence is sufficient for characterizing the shifted population. Even with small differences and many cells still in the area that would be defined as „negative" by 99% threshold gating, all cells of a shifted symmetrical population are positive. The problem is to recognize a shifted population as such and to refrain from setting analysis gates which always give wrong results.

The nonsymmetrical fluorescence distribution (Fig. 6) is most difficult to evaluate because a correct description would require curve fitting, which is not an option of most state-of-the-art software. Exceptions are the flow cytometry softwares Modfit (Verity) and Multicycle (Partec). A simplified version of curve fitting is subtraction of the histogram of the negative control, but this can be grossly misleading if the two histograms are not normed for the peak of the negative cells. In any case, interpretation of nonsymmetrical distributions must consider the possibilities of distinct but overlapping populations versus kinetic transitions.

In daily routine, for nonsymmetrical histograms with two peaks, it is assumed that these histograms result from two overlapping populations. As long as these subpopulations have about the same size, an approximation of frequency and mean fluorescence can be obtained, setting a threshold between the two peaks, where the differential of the curve reaches zero, and assuming that as many positive cells fall into the negative fraction as vice versa (valley method). This method is not suitable if one of the populations is very small, i.e., for rare cells, but it is the best that can be used with most present software.

In any case, nonsymmetrical histograms reflect a lack of resolution of immunofluorescence. In practical terms, it is worth devoting some attention to improving the resolution of staining rather than statistics.

3.4.4 From Overlapping to Separate Populations

The difficulties in evaluation of overlapping fluorescence distributions should be one of the major reasons for improving the staining reagents and protocols, as discussed above, for example, absorbing cross-reactive antibodies out of staining reagents, titrate reagents, and using indirect rather than direct staining. Higher resolution can also be obtained by changing the fluorochrome conjugated to the staining antibody, for example, from fluorescein to phycoerythrin, which has a higher quantum yield of fluorescence. Thus, the use of biotinylated antibody and fluorochrome-conjugated streptavidin is a most versatile staining system for weak cellular parameters since it combines amplification due to indirect staining with the option of exchanging fluorochromes easily.

Finally, if available, additional parameters that discriminate even weakly between the two overlapping populations should be used to improve resolution. Populations that overlap in each of two parameters may be separated in two dimensions, such as lymphocytes and monocytes from peripheral blood in forward and side scatter.

References

1. Handbook of experimental immunology, 4th edition eds. Weir, D.M., Herzenberg, L.A., Blackwell,C. and Herzenberg, L.A. Blackwell Sci. Publ., Oxford   1986
2. Clinical applications of flow cytometry: quality assurance and immunophenotyping of peripheral blood lymphocytes eds. A. Landay et al. Natl. Comm. Clin. Lab. Stand. Vol. 9, No. 13 1989
3. Methods in cell biology, Vol. 33: Flow cytometry. eds. Darzynkiewicz, Z. and Crissman, H.A. Academic Press, New York   1990
4. Logarithmic amplifiers. Gandler, W. and Shapiro, H. Cytometry 11:447-450 (1990)

4 Multicolor Immunofluorescence Analysis

D. RECKTENWALD

4.1 Background

For many research questions it is desirable to measure subsets of cell populations that have been identified by immunofluorescence. In principle this could be achieved by sorting positive and negative populations of the cells of interest, dissociating and washing out the first reagent, and then staining the two populations with another antibody reagent labeled with the same fluorophor for subsequent analysis. However, this approach is rarely taken because the availability of many different dyes suitable as labels for immunofluorescence makes the simultaneous measurement of many subpopulations in one sample possible. The first simultaneous two-color/one-laser immunofluorescence method for flow cytometry used fluorescein isothiocyanate (FITC) and rhodamine conjugates with suboptimal 514-nm excitation from an argon ion laser (Loken et al. 1977). Subsequently, FITC and Texas red conjugates were used with two-laser excitation (e.g., Titus et al. 1982). The development of phycoerythrin (PE) antibody conjugates made two-color immunofluorescence (FITC/PE) with single-wavelength excitation at 488 nm a routine method (Oi et al. 1982; Glazer et al. 1990). The more recent development of energy transfer complex conjugates, using PEs as donor and phycobiliproteins or synthetic dyes as acceptors (Glazer et al. 1983; Recktenwald et al. 1991; Ernst et al. in prep.) and the discovery that peridinin chlorophyll (PerCP) proteins are stable when conjugated to immunoglobulins and show acceptable nonspecific binding behavior (Recktenwald 1990; Recktenwald et al. 1991) allow three-color immunofluorescence analysis without significant technical difficulties today. A comprehensive list of dyes for multicolor immunofluorescence can be found in the references (Recktenwald et al. 1991; Lanier et al. 1991). Methods that combine immunofluorescence with DNA staining are not reviewed here. An example is described in Rabinovitch et al. 1986.

4.2 Material

Cells Cell suspension (e.g., lysed blood or mononuclear cells) in phosphate buffered saline (PBS) with 0.1% NaN_3, 2% fetal calf serum

Reagents – Titered solutions of antibodies (see Chap. 3) labeled with protein or energy transfer complex such as PE-Texas red or PE-CY5. Fluorochrome conjugates can be prepared as described in Sect. 2.1. Phycobiliproteins can be obtained from Molecular Probes or Cyanotech; the PerCP protein is available from Cyanotech or Advanced Algal. FITC and PE antibody conjugates can also be obtained from most companies that provide antibodies to cell surface molecules (for a list see the 1992 Biotech Buyers' Guide from the American Chemical Society, Washington D.C., U.S.A.). PerCP protein antibody conjugates are available from Becton-Dickinson and energy transfer complex antibody conjugates from Becton-Dickinson, Coulter, Gibco, and Pharmingen.
 – Laser dye LDS 751 from Exciton, Dayton, OH, USA
 – Buffer: filtered PBS with 0.1% NaN_3 and 2% fetal calf serum pH 7.2.
 – Fixative: 0.5% paraformaldehyde in PBS pH 7.2

Equipment – 12 x 75 mm test tubes or 96-well microtiter plates
 – Micropipettor 0–200 µl with tips
 – Ice bath or refrigerator
 – Vortex
 – Low-speed centrifuge for test tubes or microtiter plates
 – Flow cytometer capable of three-color fluorescence detection

4.3 Method

The method given here was revised from Monoclonal Antibodies and Source Book, Becton-Dickinson, San Jose, CA, USA.

1. Add an appropriate number of cells (typically 10^5–10^6) in a small volume of medium (i.e., 50 µl) to a mixture containing the various antibodies in about 50 µl, so that the concentration of antibodies in the final volume of 100 µl is the titered optimal concentration for each antibody.
2. Mix.
3. Incubate in the dark (many fluorescent dyes are destroyed by light) on ice or in a refrigerator (2°–8°C) for 15–30 min.
4. Centrifuge at 200 g for 5 min at 2°–8°C.
5. Carefully remove the supernatant
6. Gently flick the cell pellet (fingertip or low-speed vortex) and resuspend the cells in buffer (200 µl for microtiter plate or 1 ml for test tubes).
7. Repeat steps 4 and 5.

Example   49

8. Resuspend the pellet in 200 µl–1 ml buffer for immediate analysis or perform step 9 for fixation
9. Add 100 µl–1.0 ml of cold 0.5% paraformaldehyde to the flicked pellet and vortex or mix immediately. Store the fixed cells at 4°C in the dark until analysis.

4.4 Example

Staining of human peripheral blood with α-CD4-FITC, α-CD8-PE, and α-CD3-PerCP is illustrated in the figure. In this example blood is stained as described above, but a lysis step is introduced into the procedure to remove the excess of erythrocytes.

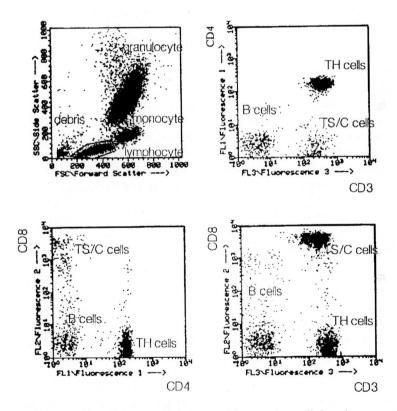

Figure 1. Multi-color immunofluorescence. Mononuclear cells from human peripheral blood were preenriched by lysis of erythrocytes (see also Chap.14) and stained as described in the text for CD3 (T cell receptor), CD4 (T helper/inducer lymphocytes, low amounts also on monocytes) and CD8 T suppressor/cytotoxic lymphocytes). The scatter and fluorescence data are shown here in various combinations as dot plots, gating the fluorescence data on lymphocytes by scatter (*upper left*). In particular combinations, additional dyes could be used even on a flow cytometer with only three fluorescence detectors, as long as different fluorescence spectra guarantee differential detection by two detectors

- 20 µl each of titered α-CD4-FITC, α-CD8-PE, and α-CD3-PerCP was mixed in a 12 x 75 mm^2 tube, 50 µl human peripheral EDTA-treated blood was added
- The sample was vortexed and incubated at room temperature for 15 min
- 2.0 ml FACS lysing solution (Becton-Dickinson, San Jose, CA, USA) was added, and the sample was vortexed again.
- After incubation for 10 min the sample was washed twice by centrifugation; the final pellet was resuspended in 0.5% formaldehyde in PBS.
- Analysis was performed with a FACScan flow cytometer.

4.5 Modifications

No removal of unbound reagent (washing) For many antibodies steps 1–3 are sufficient to obtain adequate staining for flow cytometric analysis (see Chap. 3). In these cases the cell suspension obtained after step 3 can be diluted with buffer to a volume adequate for flow analysis and measured.

Discrimination of nucleated from non-nucleated cells By omitting the far red label from the staining mixture and using FITC- and PE-labeled antibodies only, addition of the vital dye LDS 751, emitting fluorescence in the far red, or an equivalent dye such as LDS 730 or LDS 798 prior to analysis allows the discrimination of nucleated from non-nucleated particles, thus facilitating analysis of unlysed whole blood or bone marrow samples (Recktenwald 1988). Use 10 µl saturated methanolic solution of the dye for 1 ml cell suspension. Alternatively, an aqueous suspension of the LDS 751 dye can be prepared by adding 10 µl 1 mg/ml solution in methanol or dimethylsulfoxide to 10 ml PBS. LDS 751 can also be used to discriminate between cells that had been damaged before fixation (Terstappen et al. 1988).

4.6 Tips, Tricks, and Troubleshooting

Instrument set-up Amplifier gains can be calibrated to a fixed sensitivity level by using stable fluorescent materials such as polymer microparticles or glutaraldehyde-fixed avian erythrocytes. Alternatively, an unstained cell sample can be used to adjust the instrument amplification to position autofluorescence at the lower end of the measurement scale. To adjust the settings for fluorescence multicolor compensation, it is best to mix samples stained with one dye each independently, to „dial out" spectral overlap (after the amplifier gains are set). Accurate compensation settings are imperative, when double-stained populations of low intensity are to be analyzed.

Most immunofluorescence measurements use light scatter as the instrument trigger signal. For the analysis of rare populations in a sample, for example, leukocyte subsets in the presence of red cells in blood, the use of immunofluorescence or DNA-related fluorescence as trigger signal can improve sample throughput and data quality. **Fluorescence instrument trigger**

Isotype controls are used to check for nonspecific binding with a particular cell preparation. A separate sample of the cell suspension is stained with conjugated antibodies without reactivity for antigens of the cell preparation of interest. Ideally the isotypes of the controls should match the isotypes of the antibodies for specific staining. A cellular control for the specific antibody is an even better control (see Chap. 3). **Controls**

Labels with bright fluorescence should be used for the antigens with the lowest density. Generally, PE and PE-based energy transfer conjugates have the brightest fluorescence. In the case of cells with high autofluorescence, allophycocyanin yields more relative brightness than PEs. The combination of PerCP and PE as labels is the ideal choice for measuring a low-density antigen (PE label) on a subpopulation defined by another antigen of medium or high density (PerCP label) because no substantial spectral compensation is needed between those two dyes. (PerCP shows zero emission below 600 nm). **Label selection**

PerCP conjugates provide their optimal sensitivity with an excitation laser power below 20 mW. For use with water-cooled high-power lasers the excitation energy should be lowered until optimal resolution for all labels is obtained.

Most fluorescent labels are light sensitive. To analyze dim populations and prevent artifacts, samples should be kept away from bright light. **Photo-bleaching**

Fluorescein

Acidic buffer conditions should be avoided during the analysis of samples stained with FITC because the fluorescence of the dye and its conjugates is pH dependent. The acidic form of the dye is nonfluorescent.

References

Glazer AN. Stryer L. Phycobiliprotein-avidin and phycobiliprotein-biotin conjugates. Methods Enzymol. 1990. 184. P 188-94.

Glazer AN. Stryer L. Fluorescent tandem phycobiliprotein conjugates. Emission wavelength shifting by energy transfer. Biophys J. 1983 Sep. 43(3). P 383-6.

Lanier LL. Recktenwald DJ. Multicolor immunofluorescence and flow cytometry Methods: a companion to methods in enzymology. 1991 June. 2(3). P 192-199

Loken MR. Parks DR. Herzenberg LA. Two-color immunofluorescence using a fluorescence-activated cell sorter. J Histochem Cytochem. 1977 Jul. 25(7). P 899-907.

Oi VT. Glazer AN. Stryer L. Fluorescent phycobiliprotein conjugates for analyses of cells and molecules. J Cell Biol. 1982 Jun. 93(3). P 981-6.

Rabinovitch PS. Torres RM. Engel D. Simultaneous cell cycle analysis and two-color surface immunofluorescence using 7-amino-actinomycin D and single laser. J Immunol. 1986 Apr 15. 136(8). P 2769-75.

Recktenwald D. Method for analysis of subpopulations of blood cells. US Patent No. 4,727,020. February 23, 1988.

Recktenwald D. Peridinin chlorophyll complex as fluorescent label. US Patent No. 4,876,190. October 24, 1989.

Recktenwald D. Prezelin B. Chen CH. Kimura J. Biological pigments as fluorescent labels for cytometry. New Technologies in Cytometry and Molecular Biology, Gary C. Salzman, Editor, Proc. SPIE 1206 P 106-111 (1990)

Terstappen LWMM. Shah VO. Conrad MP. Recktenwald D. Loken MR. Discriminating between damaged and intact cells in fixed flow cytometric samples. Cytometry 1988 September. 9(5). P 477-484.

Titus JA. Haugland R. Sharrow SO. Segal DM. Texas red, a hydrophilic, red-emitting fluorophore for use with fluorescein in dual parameter flow microfluorometric and fluorescence microscopic studies. J Immunol Methods. 1982. 50(2). P 193-204.

5 Combined Intracellular and Surface Staining

M. ASSENMACHER

5.1 Background

Detection of intracellular proteins by immunofluorescence allows one to determine the frequency and the light scatter and surface immuno-phenotype of protein-producing cells, irrespective of whether the protein is to be secreted, membrane bound, or localized in the cytoplasm. This is the only way to analyze secreted, cytoplasmic, and nuclear proteins by flow cytometry. Unfortunately, to date no method is known for cytoplasmatic immunofluorescence of live cells. In order to allow staining antibodies to penetrate the cell membrane, cells must be fixed and the membranes permeabilized. The choice of fixation method depends on the protein and its intracellular location and on the further use of the cells to be analyzed. For several applications fixation in formaldehyde and permeabilization of cell membranes by saponin (Willingham, M.C. and I. Pastan 1985) has been used successfully, including assessment of cytokines (Sander et al. 1991, Schmitz et al., 1992, in press). Formaldehyde is a cross-linking fixative with good preservation of cell morphology (Williams and Chase 1976; see figure, below). The plant glycoside saponin, a mild nonionic detergent, complexes with membrane cholesterol and other unconjugated β-hydroxy-sterols leading to ring-shaped complexes with a central pore about 8 nm in diameter (Bangham and Horne 1962; Glauert et al. 1962). These pores allow passage of molecules of up to 200 kDa, (Schulz 1990). Since saponin acts in a reversible way (Willingham and Pastan 1985), it must be present in all incubation and washing steps; however, it allows surface staining at any point of the procedure if removed from the staining solution.

Alternatives are the use of either other detergents, such as NP40 and digitonin in combination with fixation by formaldehyde, or of other organic solvents, for example, 70% methanol or ethanol/acetic acid (95/5), which fix and permeabilize cells in one step. Especially for staining of RNA and DNA, fixation with alcohol is preferable, not only because the staining works better, but also because DNA and RNA are less degraded and can be used for molecular analysis of sorted cells.

Since staining antibodies must penetrate the cells and diffuse through the cytoplasm, much longer times are required for staining and washing. Due to the long incubation times, which support low-affinity cross-reactions, and to the overall stickiness of cytoplasm, background staining is a common problem. Thus high purity of reagents and careful titration of staining parameters become extremely important (see Chap. 3). Absorption of polyclonal antibodies on liver powder (acetone precipitate of liver) or on irrelevant cells [2:1, volume of antibody solution (1 mg/ml) : packed cells] often reduces unspecific staining to an acceptable level.

A second problem with intracellular staining is that, after fixation, discrimination between cells that had been alive before fixation and those that were dead becomes difficult. This may disturb the analysis due to unspecific fluorescence from the necrotic/apoptotic cells. Such cells can be removed prior to fixation by Ficoll gradient centrifugation (see Chap. 14) or they can be gated out after staining them with LDS 751 (see Chap. 4, Terstappen et al. 1988).

5.2 Material

Cells
- BALB/c mouse spleen cells stimulated with 2 µg/ml Staphylococcus aureus enterotoxin B (SEB; Sigma, St. Louis, MO) for 44 h at 2×10^6 cells/ml.
- As high controls X63Ag8.653 myeloma transfectants constitutively expressing murine interleukin (IL) 2 or IL-5 cDNA (Karasuyama and Melchers 1988) and 3T3 fibroblasts transfected with murine IL-4 cDNA (Müller 1987) were used.

Reagents
- **Antibodies:**
  Rat anti-mouse cytokine monoclonal antibodies (mAbs)
  Anti-mouse IL-2 S4B6 (Mosmann et al. 1986)
  Anti-mouse IL-4 11B11 (Ohara and Paul 1985)
  Anti-mouse IL-5 TRFK5 (Schumacher et al. 1988)
  Anti-mouse interferon (IFN)-γ R46A2 (Spitalny and Havell 1984)
  Goat anti-rat IgG fluorescein (FITC; Southern Biotechnology Associates, Birmingham, AL) (reabsorbed on µ-, κ- immunoglobulin (Ig) and mouse spleen cells)
  Anti-mouse CD4 GK-1.5 (Dialynas et al. 1983) conjugated to phycoerythrin

- **Buffer:**
- Phosphate-buffered saline (PBS)
- PBS with 0.5% bovine serum albumin (BSA) + 0.02% sodium azide
- $(PBS/BSA/NaN_3)$
- $PBS/BSA/NaN_3$ with 0.5% saponin (Sigma; saponin buffer)

- **Fixative:**
- 4% Formaldehyde (Merck, Darmstadt, FRG) in PBS

Example   55

5.3 Method

1. Wash cells once with PBS.
2. Resuspend cells in PBS at 2×10^6 /ml.
3. Add 1 vol 4% formaldehyde/PBS for 20 min at room temperature.
4. Wash twice with PBS.
5. Resuspend in PBS/BSA/NaN$_3$ at $1–2 \times 10^6$ cells/ml.
6. Store at 4°C in the dark until staining.

1. Use about 1.5×10^6 cells for each 1.5-ml test tube sample.
2. Spin down for 10 min at 350 g.
3. Incubate pellet with 30 µl rat anti-mouse cytokine mAb (titrated concentration) in saponin buffer for 60 min at room temperature.
4. Fill tube up with saponin buffer and spin down for 10 min at 350 g.
5. Wash with 1 ml saponin buffer for 30 min at room temperature.
6. Spin down for 10 min 350 g.
7. Incubate pellet with 30 µl goat anti-rat IgG FITC in saponin buffer for 45 min at room temperature.
8. Fill up and spin down for 10 min at 350 g.
9. Wash with 1 ml saponin buffer for 30 min at room temperature.
10. Spin down for 10 min at 350 g.
11. Resuspend in PBS/BSA/NaN$_3$.

1. Spin down for 10 min at 350 g.
2. Incubate pellet with 30 µl Ab solution (e.g.: a-CD4 GK-1.5-PE, titrated) in PBS/BSA/NaN$_3$ for 15 min at room temperature.
3. Fill up and spin down for 10 min at 350 g.
4. Resuspend in about 500 µl PBS/BSA/NaN$_3$.

5.4 Example

Cytoplasmic expression of cytokines in correlation with surface immunophenotype has been analyzed in mouse spleen cells stimulated in vitro with the superantigen SEB. After 44 h of stimulation, cells were fixed and stained for IL-2, IL-4, IL-5 and IFN-γ in the cytoplasm and for CD4 on the surface, as described below. Almost no IL-4 or IL-5 producing cells were detected either by fluorescence microscopy or by flow cytometry. In the microscope, the characteristic staining pattern for IL-2 and IFN-γ could be observed, local perinuclear immunofluorescence (Fig. 1), reflecting accumulation of cytokines in the Golgi compartment (Sander et al. 1991). For FACS analysis, large activated lymphocytes were gated according to light scatter and autofluorescence (Fig. 2). IL-2 and IFN-γ producing cells were analyzed in correlation to CD4 expression (Fig. 3). SEB-stimulated T cell blasts produce varying amounts of cytokines resulting in a wide distribution of fluorescence intensity (Fig. 3).

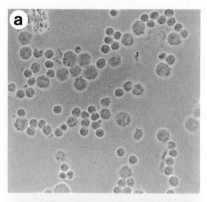

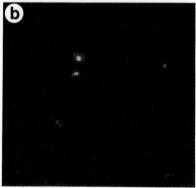

Figure 1. Photographs of intracellular immunofluorescence of IFN-γ staining. Murine spleen cells were stimulated for 42 h with SEB and stained for IFN-γ as described. a Phase contrast. b Fluorescent staining of IFN-γ in the same cells. Note the local, perinuclear spots of fluorescence and the large size of positive cells. (Zeiss Axiophot equipment and Kodak Tmax 400 film)

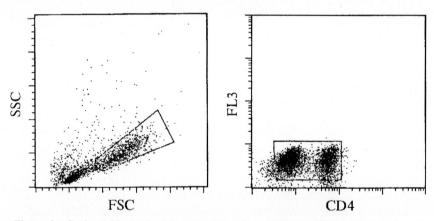

Figure 2. Gating of large lymphocytes (blasts) according to light scatter and autofluorescence properties. Dot plots display forward and side scatter at linear amplification and fluorescence at logarithmic amplification. [A FACScan and FACScan research software (Becton-Dickinson) were used for flow cytometry analysis.]

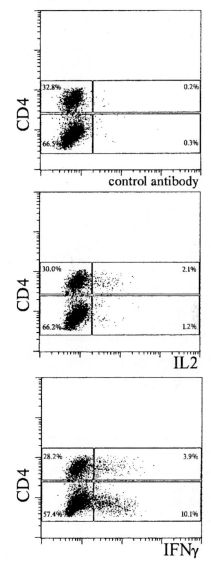

Figure 3. Determination of frequency and immunophenotype of IL-2 and IFN-γ producing blasts. Mouse spleen cells stimulated with SEB for 44 h were stained intracellularly for IL-2 and IFN-γ and for CD4 on the surface as described. Large lymphocytes (blasts) were gated as shown in previous figure. Data on 10 000 blasts per sample were collected and analyzed using FACScan research software

5.5 Tips, Tricks, and Troubleshooting

Staining too weak
- Use control of transfected cells.
- Use microscope to control for localization of staining.
- Use indirect staining with polyclonal antibodies to amplify signals.

High background
- Absorb X reactions on affinity sorbents.

– Absorb on cells:

1:1 or 2:1, volume of staining solution stock (<1mg/l) : packed, live cells of irrelevant surface phenotype, mix gently, incubate on ice for >1 h.

Spin down for 10 min at 350 g, recover supernatant, spin down in Eppendorf centrifuge to remove aggragates, titrate out, store at -70^{0}C in aliquots until use.

References

– Bangham, A.D., and R.W. Horne. 1962. Action of saponin on biological cell membranes. Nature 196:952

– Dialynas, D.P., Wilde, D.B., Marrack, P., Pierres, A., Wall, K.A., Havran, W., Otten, G., Loken, M.R., Pierres, M., Kappler, J., and F.W. Fitch. 1983. Characterization of the murine antigenic determinant, designated L3T4a, recognized by monoclonal antibody GK1.5: expression of L3T4a by functional T cell clones appears to correlate primarily with class II MHC antigen reactivity. Immunol. Rev. 74:29.

– Glauert, A.M., Dingle, J.T., and J.A. Lucy. 1962. Action of saponin on biological cell membranes. Nature 196:953

– Karasuyama, H., and F. Melchers. 1988. Establishment of mouse cell lines which constitutively secrete large quantities of interleukin 2,3,4 or 5 using modified cDNA expression vectors. Eur. J. Immunol. 18:97.

– Mosmann, T.R., Cherwinski, H., Bond, M.W., Giedlin, M.A., and R.L. Coffmann. 1986. Two types of murine helper T cell clones. I. Definition according to profiles of lymphokine activities and secreted proteins. J. Immunol. 136:2348.

– Müller, W., Ph.D. thesis. University of Cologne. 1987

– Ohara, J., and W.E. Paul. 1985. Production of a monoclonal antibody to and molecular characterization of B cell stimulatory factor 1. Nature 315:333.

– Sander, B., Andersson, J., and U. Andersson. 1991. Assessment of cytokines by immunofluorescence and the paraformaldehyde-saponin procedure. Immunol. Rev. 119:65

– Schmitz, J., Assenmacher, H., and A. Radbruch. Regulation of Th cell cytokine expression: Functional dicrotomy of APCs. Eur. J. Immunol. in press.

– Schulz, I. 1990. Permeabilizing cells: some methods and applications for the study of intracellular processes. Methods-Enzymol. 192:280.

– Schumacher, J.H., O'Garra, A., Shrader, B., van Kimmenade, A., Bond, M.W., Mosmann, T.R., and R.L. Coffmann. 1988. The characterization of four monoclonal antibodies specific for mouse IL5 and development of mouse and human IL5 enzyme linked immunosorbent assays. J. Immunol. 141:1576.

– Spitalny, G.L., and E.A. Havell. 1984. Monoclonal antibody to murine gamma interferon inhibits lymphokine-induced antiviral and macrophage tumoricidal activities. J. Exp. Med. 159:1560.

– Terstappen, L.W.M.M., Shah, V.O., Conrad, M.P., Recktenwald, D. and M.R. loken. 1988. Discriminating between damaged and intact cells in fixed flow cytometric samples. Cytometry 9:477.

– Williams, C.A., and M.W. Chase. 1976. Methods in immunology and immunochemistry. vol. V. Academic Press.

– Willingham, M.C., and I. Pastan. 1985. An atlas of immunofluorescence in cultured cells. vol. II. Academic Press.

6 Scatchard Analysis by Flow Cytometry

R.F. MURPHY

6.1 Background

The goal of the protocols described here is to measure the affinity of a ligand for cell surface receptor(s) and to determine the number of receptors per cell, following the classic method of Scatchard [1]. The preferred method for analysis of ligand binding is to measure ligand bound to cells at equilibrium without removing unbound ligand, since removing unbound ligand allows ligand dissociation to begin and may result in underestimation of the equilibrium value. As first described by Bohn [2] and Steinkamp and Kraemer [3], the small diameters of the sample stream and excitation beam used in most flow cytometers provide excellent discrimination between cell-associated and free ligand. Further discussion of the basis of this discrimination and analysis of the effects of ligand affinity, number of receptors per cell, and sample stream diameter on expected signal (fluorescence from bound ligand) to noise (fluorescence from free ligand) may be found elsewhere [4]. When the equilibrium method (Sect. 6.3.1) is not suitable, the rapid dilution method (Sect. 6.3.2) is frequently an acceptable substitute.

6.2 Material

Fluorescent ligand conjugates

Ligand conjugates may be purchased (e.g., Molecular Probes, Eugene, OR) or prepared following standard procedures. For illustration, a lissamine rhodamine sulfonyl chloride (LRSC) conjugate of transferrin (Tf) is used below (see [5] for preparation of Tf conjugates). Stock solutions of LRSC Tf at 1 mg/ml are normally prepared. For the protocols below, a 10 mg/ml stock solution of unconjugated diferric human transferrin in phosphate-buffered saline (PBS) is also needed to confirm the specifity of the fluorescent conjugates.

Miscellaneous

PBS containing 8 mM NaH$_2$PO$_4$, 2.7 mM KCl, 140 mM NaCl, and 1.5 mM KH$_2$PO$_4$ (adjusted to pH 7.4) is used for washing and labeling cells. The appropriate base salt solution for a given cell type [e.g., Minimum essential medium/Alpha Mod (α-MEM) salts] may be substituted. Growth medium should not be used since many components of typical media (e.g., flavins, phenol red) increase background fluorescence and may quench fluorescence.

6.3 Methods

The concentrations of Tf presented below are for illustration; appropriate concentrations for other ligands should be determined based on published affinities, or a wide range of concentrations used initially to find concentrations which result in measurable specific binding.

6.3.1 Equilibrium Scatchard Analysis

1. Collect at least 5×10^7 cells, wash twice with PBS at room temperature (these washes should dissociate unlabeled ligand on surface receptors). Resuspend cells in PBS at 10^7 cells/ml on ice.
2. Take 0.25 ml aliquots of cells and add to flow cytometer sample tubes containing 0.25 ml PBS with various concentrations of LRSC Tf and unlabeled Tf, as follows: 0, 0.067, 0.2, 0.67, 2 µg/ml LRSC Tf, four tubes for each concentration, two with („blocked") and two without („unblocked") 1 mg/ml unlabeled Tf. Incubate on ice for 60 min.
3. Analyze by flow cytometry using minimum sample volume flow rate (for stream-in-air flow cytometers this minimizes the core diameter) and keeping samples at $0°–4°C$ during analysis. At low sample volume flow rates, there may be an appreciable delay before events are detected. If a „boost" is used to shorten this delay, make sure that the fluidics has equilibrated (i.e., that the core diameter is stable) before beginning data acquisition. This is most easily done by waiting for the event rate to stabilize.
4. Calculate specific binding as mean fluorescence for unblocked samples minus mean fluorescence for blocked samples for each concentration of LRSC Tf. Display specific binding/LRSC concentration versus specific binding (bound/free versus bound) according to the method of Scatchard [1] to determine K_d and maximum amount of binding. The latter value (in fluorescence channels) may be converted to number of receptors per cell using calibration particles. Be sure to take into account the number of dyes per ligand when performing this conversion.

6.3.2 Fast Nonequilibrium Scatchard Analysis

1. Collect at least 5×10^7 cells, wash twice with PBS at room temperature (these washes should dissociate unlabeled ligand on surface receptors). Resuspend cells in PBS at 5×10^7 cells/ml on ice.

Example   61

2. Take 20 µl aliquots of cells and add to flow cytometer sample tubes containing 20 µl PBS with various concentrations of LRSC Tf and unlabeled Tf, as follows: 0, 0.2, 0.67, 2, 6.7 µg/ml LRSC Tf, four tubes for each concentration, two with („blocked") and two without („unblocked") 1 mg/ml unlabeled Tf. Incubate on ice for 60 min.
3. Add 1 ml cold PBS to each tube immediately before analyzing. Keep sample on ice while on flow cytometer.
4. Calculate specific binding as mean fluorescence for unblocked samples minus mean fluorescence for blocked samples for each concentration of LRSC Tf.

6.4 Example

The figure shows example histograms obtained using protocol 6.3.1. Note that binding is easily detectable at both ligand concentrations, but that nonspecific fluorescence increases in panel B. This nonspecific fluorescence is due to a combination of LRSC Tf nonspecifically bound to cells and fluorescence from unbound LRSC Tf. The latter component depends on sample core diameter, a function of sample volume flow rate. In duplicate samples, the average ratio of blocked to unlabeled („0") increased from $1.26 + 0.07$ to $2.00 + 0.02$, and the nonspecific fluorescence (blocked) as a percentage of total fluorescence (unblocked) increased from 3% to 5%. Even this percentage of nonspecific fluorescence may be acceptable for some applications (e.g., screening or sorting for cells with high or low receptor numbers).

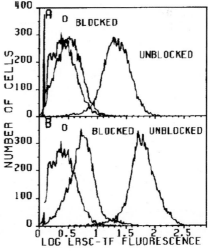

Figure 1. Detection of transferrin binding at equilibrium by flow cytometry. K562 cells were incubated with 0.67 µg/ml (A) or 2 µg/ml (B) LRSC Tf in the presence („blocked") or absence („unblocked") of 1 mg/ml unlabeled Tf according to protocol 6.3.1. Cells receiving no LRSC Tf were also analyzed (0).

References

1. Scatchard, G (1949) Ann. N.Y. Acad. Sci. 51, 660
2. Bohn, B (1976) Exp. Cell Res. 103, 39-46
3. Steinkamp, JA, and Kraemer, PM (1979) In: Flow cytometry and sorting. Melamed MR, Mullaney PF, Mendelsohn ML (eds.), John Wiley and Sons, New York, pp. 497-504
4. Murphy, RF (1990) In: Flow cytometry and sorting, Second Edition. (Melamed, MR, Lindmo T, and Mendelsohn, ML, eds.), Wiley-Liss, Inc., New York, pp. 355-366
5. Sipe, DM, and Murphy, RF (1987) Proc. Natl. Acad. Sci. U.S.A. 84, 7119-7123

Part III DNA and Proliferation

7 Preparation and Staining of Cells for High-Resolution DNA Analysis

F.J. OTTO

7.1 Background

The extreme constancy of the cellular DNA content of tissues and clonal cell populations presents a challenge for preparation and staining methods as well as for measuring techniques. In recent years, flow cytometry has proven to be particularly suited for high-resolution DNA analysis. Protocols for appropriate preparation and staining have been developed and published by several groups. Good results have been obtained by the methods of Vindelov [1] and Thornthwaite [2], for instance.

In 1981 we presented a method for producing DNA histograms with small, reproducible coefficients of variation that made them applicable in mutagenicity studies [3]. This method proved useful in many fields of biological and medical research as well as routine work [4, 5]. It is based on treatment with the detergent Tween 20 in citric acid solution to disperse solid tissues and to produce a suspension of cell nuclei fixed with ethanol and stained with 4',6-diamidino-2-phenylindole (DAPI). DAPI was introduced in flow cytometry by Göhde et al. [6] and has proven favorable because of its high specificity and fluorescence intensity.

7.2 Material

- Detergent solution: 100 ml deionized water, 2.1 g citric acid x H_2O (= 0.1 mol/l), 0.5 g Tween 20 (Serva 37470).
- Staining solution: 100 ml deionized water, 7.1 g Na_2HPO_4 x $2H_2O$ (= 0.4 mol/l), 0.2 mg DAPI (Partec 7202); store in the dark at room temperature. These solutions are stable at room temperature for 1–2 weeks.

7.3 Method

1. Mince tissue with scissors or scalpels.
2. Suspend in appropriate volume of detergent solution.
3. Incubate at room temperature for 20 min while shaking gently.

4. Put the suspension through a 50–100 µm mesh sieve.
5. Centrifuge 10 min at 100 *g*.
6. Remove supernatant.
7. Resuspend cell pellet in a small volume of phosphate-buffered saline.
8. Fix with 70% ethanol.
9. Store for a few days.
10. Centrifuge 10 min at 200 *g*.
11. Remove fixative completely.
12. Resuspend cell pellet in 1 volume (e.g. 1 ml) of detergent solution.
13. Incubate at room temperature for 10 min while shaking.
14. Add 5 volumes (e.g. 5 ml) of staining solution.
15. Keep in dark at room temperature for some hours or over night.
16. Analyze by flow cytometry.

7.4 Example

This protocol yields highly resolved DNA histograms from a large variety of tissues. For example, a histogram of a primary tumor of malignant melanoma is shown in Figure 1, exhibiting a peak of normal diploid cells (at channel 100) and two aneuploid tumor cell lines.

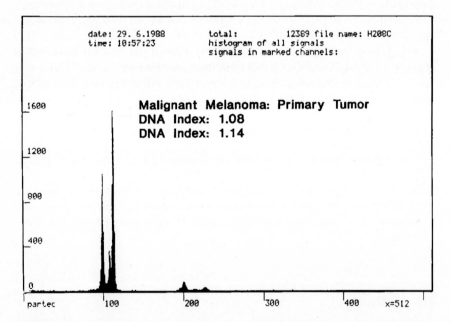

Figure 1. Flow cytometric DNA histogram of a primary malignant melanoma biopsy. Using this protocol, differences in cellular DNA content down to 2% can be safely detected. In tumor samples this resolution is thought to be indispensable for reliable assessment of aneuploidies

7.5 Modifications

With cells that are already in suspension, such as lymphocytes, fixation in 70% ethanol is possible without previous detergent treatment. For the detergent treatment after fixation a double concentrated citric acid solution (= 0.2 mol/l) should be used in these cells.

7.6 Tips, Tricks, and Troubleshooting

Samples should be fixed for at least 3–4 days. These fixed cells can be stored for up to several months; however, the time of fixation may influence the stainability to some extent. Therefore, if reference cells are used for calibration and establishing of DNA indices, they should be fixed at the same time.

Fixatives others than ethanol are not recommended. Especially formalin fixation is not suitable for subsequent DAPI staining.

It is important to keep the staining solution and the stained cells at room temperature. Lower temperatures cause the highly concentrated sodium hydrogen phosphate to form crystals which can be dissolved only slowly, and which may damage the stained cells.

The fluorochrome DAPI obviously needs some time for quantitative binding to nuclear DNA. As a consequence and in order to obtain optimal results when measuring, stained cells should be kept a few hours or over night.

It should be pointed out that DAPI preferentially stains AT-rich DNA. Therefore it is not suitable for determining absolute amounts of DNA or for comparing cells with different AT contents. Although certain restrictions in use must be observed, it is possible to use chicken or trout erythrocytes as standards to determine the relative peak positions or the coefficients of variation of particular cell populations.

References

1. Vindelov, L.L., Christensen, I.J. and Nissen, N.I.: A detergent-trypsin method for the preparation of nuclei for flow cytometric DNA analysis. Cytometry 3: 323-327, 1983.
2. Thornthwaite, J.T., Sugarbaker, E.V. and Temple, W.J.: Preparation of tissues for DNA flow cytometric analysis. Cytometry 1: 229-237, 1980.
3. Otto, F.J., Oldiges, H., Göhde, W. and Jain, V.K.: Flow cytometric measurement of nuclear DNA content variations as a potential in vivo mutagenicity test. Cytometry 2: 189-191, 1981.
4. Otto, F.: DAPI staining of fixed cells for high-resolution flow cytometry of nuclear DNA. In: Darzynkiewicz, Z. and Crissman, H.A. eds.: Methods in cell biology, Vol. 33, San Diego, pp. 105-110, 1990.
5. Otto, F.J., Schumann, J. and Bartkowiak, D.: High-resolution DNA flow cytometry in malignant melanoma. Cytometry Supplement 4: 55, 1990.
6. Göhde, W., Schumann, J. and Zante, J.: The use of DAPI in pulse cytophotometry. In: Lutz, D. ed.: Pulse cytophotometry, Ghent, pp. 229-232, 1978.

8 Simultaneous Flow Cytometric Detection of Bromodeoxyuridine Incorporation and Cell Surface Marker Expression

W. MÜLLER

8.1 Background

Bromodesoxyuridine (BrdU) is an analog to thymidine. It can be incorporated with high efficiency into DNA during DNA synthesis by the cell-replacing thymidine residues. A large panel of monoclonal antibodies to BrdU were develop which are able to bind to DNA molecules containing BrdU [1-3]. For the detection of BrdU-labeled DNA by antibodies only a small number of BrdU molecules are needed. This grade of substitution can be reached at BrdU concentrations which are nontoxic for cells. By adding BrdU to the drinking water it is even possible to deliver BrdU constantly to mice over long period (several months) [5,6]. By combining cell surface staining with anti BrdU-antibody staining it is possible to analyze life spans of cell populations in mice [4-7]. In cell cultures it can be used to determine the percentage of cells proliferating at a given time. In combination with DNA stains it furnishes additional resolution to cell cycle analysis.

Maximum staining intensities by the anti-BrdU antibody is achieved after even a short labeling period. Therefore it is impossible to distinguish between cells which have gone through one or several cell divisions (see Chap. 9).

8.2 Material

- Cooling centrifuge, ice 4 °C, water bath 37 °C
- BrdU (Sigma B5002) **Reagents**
- Phosphate-buffered saline (PBS)
- PBS/1% BSA/0.03% NaN$_3$
- 1 N HCl/ 0.5% Tween 20 (prepare fresh immediately prior to use)
- 0.1 M disodium tetraborate, pH 9.0
- PBS/1% paraformaldehyde (prepare fresh)
- 70% ethanol in water
- Anti-BrdU antibodies: The protocols shown here represent optimized **Antibodies** procedures for two anti-BrdU antibodies. (Only antibody preparations kindly provided by Dr. Ternynck and Dr. Katzman were used. The

antibody preparations commercially available were not tested). Many other anti BrdU antibodies are known. These probably also work in similar procedures. All steps used for the detection of BrdU must be optimized when a different anti-BrdU antibody is used.

- Use of antibody 76-7 (Porstmann et al., 1985) results in a high resolution in the BrdU staining but requires denaturation of the cellular DNA by HCl treatment. Cell surface staining is therefore limited to HCl-resistant fluorescence dyes (e.g., fluorescein isothiocyanate, FITC). This antibody is commercially available from Immunotech.

- Antibody BU-1 (Gonchoroff et al., 1985, 1986) binds to BrdU-containing DNA without denaturation by HCl. Permeablization of the cells is performed by 70% ethanol. This allows surface staining with non HCl-resistant fluorescence dyes (e.g., phycoerythrin) but gives lower resolution of the BrdU staining compared to the antibody 76-7. The antibody BU-1 is commercially available from IBL Research Products.

Second-step reagents
- Anti-IgG1(a) Ig(4a)10.9, biotin conjugate (Oi and Herzenberg, 1979) (available from Pharmingen, cat. no. 05002D; needed for the 76-6 antibody).
- Anti IgG2a(a) Ig(1a), biotin conjugate or FITC conjugate (Pharmingen, cat. no. 05022D, biotin; or cat. no. 05034D, FITC; needed for the BU-1 antibody).
- Streptavidin phycoerythrin (Becton-Dickinson, cat. no. 9023).
- Antibodies for cell surface staining (each antibody conjugate must be tested in the fixation procedure.)

8.3 Method

BrdU labeling
Cells are labeled either in vitro or in vivo for the desired length of time. In the case of in vivo labeling mice are fed with BrdU (Sigma B5002) at a concentration of 1 mg/ml in drinking water. (BALB/c mice tolerate BrdU feeding over several months; C57Bl/6 mice can be fed only up to 1 week). BrdU is light sensitive; therefore, the bottles must be wrapped with foil to protect the water from light. BrdU is a mutagen. Be careful to avoid contact to BrdU powder and water containing BrdU.

Cell surface staining
1. Prepare cell suspensions (depletion of erythrocytes is not necessary). All cells should be viable. If necessary, remove dead cells by Ficoll gradient or cotton wool column (see Chap. 14).
2. Use 1×10^7 cells/sample. Many cells become lost during the washing steps in the staining procedure. To compensate for this cell loss one must start with high cell numbers.
3. Spin down cells in an Eppendorf tube (120 g), remove the supernatant, and stain the cells in 15 µl FITC-conjugated antibody (titrated concentration) for 20 min on ice.

4. Wash cells once in 1 ml cold PBS (no protein!), spin down, and resuspend in 200 µl PBS.

For each sample use a 15-ml centrifugation tube filled with 5 ml 70% **Fixation** ethanol. To avoid cell clumping in the fixation step the cell suspension is injected into the ethanol using either an Eppendorf pipette or 1-ml syringe and small needle and is further mixed with the ethanol by gentle shaking of the tube. Incubate for 30 min on ice.

1. *DNA denaturation:*
– Spin down the cells at 400 g (from now on always use 400 g as the cell **Anti-BrdU** density changes during the fixation). During the centrifugation step **staining with** prepare the 1 N HCl/0.5% Tween 20 solution. **antibody 76-7**
– Remove ethanol, wash once in PBS, resuspend the cells in 1 N HCl/0.5% Tween 20, and incubate for 15 min at 37^{0}C.

2. *Neutralization:*
– Spin down cells, remove supernatant, and resuspend in 200 µl 0.1 M disodium tetraborate, pH 9.0, and fill up with 5 ml ice-cold PBS/ 1%BSA/0.03% NaN$_3$.
  Spin down, remove supernatant, resuspend in 200 µl PBS/BSA/NaN$_3$, and transfer to Eppendorf tubes.

3. *Staining:*
– From now on all steps must be performed on ice! Washing must be performed carefully. Allow the cells to incubate in the washing solution for 5 min before centrifugation. This is necessary to allow enough time for diffusion of the nonbound antibody out of the cells.
– Spin down the cells, remove the supernatant, and incubate with 10 µl anti-BrdU (0.1 mg/ml) antibody for 20 min.
– Wash with 1 ml PBS/1% BSA/0.03% NaN$_3$.
– Incubate with 10 µl anti IgG1a biotin (0.1 mg/ml) for 20 min.
– Wash with 1 ml PBS/1% BSA/0.03% NaN$_3$. Stain with 10 µl streptavidin phycoerythrin (e.g., Becton-Dickinson (BD) 1:2 diluted in PBS/1% BSA/0.03% NaN$_3$) for 20 min.
– Wash with 1 ml PBS/1% BSA/0.03% NaN$_3$ and resuspend the cells in 200 µl PBS/1% BSA/0.03% NaN$_3$.
– Add 200 µl icecold PBS/1% paraformaldehyde for fixation, mix, and allow to fix at least 5 min before analysis.

1. Spin down the cells at 400 g (from now on always use 400 g as the cell **Anti-BrdU** density changes during the fixation). **staining with**
2. Remove ethanol and wash the cells once in PBS/1% BSA/0.03% NaN$_3$. **antibody** Incubate cells with 100 µl BU-1 (0.3 mg/ml) at room temperature for 30 **BU-1** min.
3. Washing must be performed carefully. Allow the cells to incubate in the washing solution for 5 min before centrifugation. This is necessary to allow enough time for diffusion of the nonbound antibody out of the cells.

4. Wash with 1 ml PBS/1% BSA/0.03% NaN$_3$.
5. Incubate with 10 µl anti-IgG2a(a) biotin (0.1 mg/ml) for 20 min.
6. Wash with 1 ml PBS/1% BSA/0.03% NaN$_3$.
7. Stain with 10 µl streptavidin phycoerythrin (e.g., BD 1:2 diluted in PBS/ 1% BSA/0.03% NaN$_3$) for 20 min.
8. Wash with 1 ml PBS/1% BSA/0.03% NaN$_3$ and resuspend the cells in 200 µl PBS/1% BSA/0.03% NaN$_3$. Or, instead of the above steps 5–8, incubate with 10 µl anti-IgG2a(a) FITC (0.1 mg/ml) for 20 min.
9. Wash with 1 ml PBS/1% BSA/0.03% NaN$_3$ and resuspend the cells in 200 µl PBS/1% BSA/0.03% NaN$_3$.

Necessary controls Always include cells which do not contain a BrdU label as a negative control. Since the staining method sometimes results in a high background, such a control is necessary to adjust the flow cytometer for the background staining of cells not labeled with BrdU.

8.4 Example

Double staining of mouse bone marrow cells with anti-CD45R (B220) and anti-BrdU antibody 76-7 is shown in Figure 1.

C57Bl/6 mice were fed with BrdU in drinking water for 3 days. Bone marrow cells were stained as using anti CD45R (B220) antibody for cell surface staining and 76-7 antibody for anti BrdU staining as described. Cells were analyzed on a FACScan, and the data were analyzed using FACScan Research Software (BD). The data are represented in a two dimensional dot plot. On the left, staining of bone marrow cells of mice fed with water only is shown (negative control). On the right, staining of bone marrow cells of mice treated with BrdU is shown. One can identify three

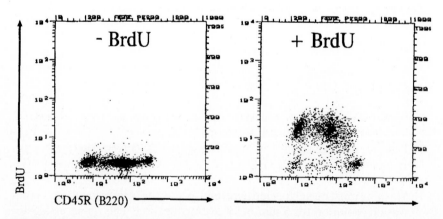

Figure 1. Combined staining of surface markers and DNA/BrdU

cell populations based on the expression levels of the CD45R (B220) marker: cells with negative, dull, and bright stain. Cells stained brightly with CD45R (B220) are not labeled by BrdU. For a discussion see [4]. For further examples of BrdU stainings see [4-7].

8.5 Tips, Tricks, and Troubleshooting

– The samples were not washed properly. The anti BrdU staining is a **Staining** nuclear staining. Is is important to wash the cells carefully. Incubate the **background** cells in the washing buffer for 5 min before centrifugation to allow **of non-BrdU** diffusion of the dye. Try to wash the samples one more time. For **labeled cells** effective washing it is also important that the cells be resuspended very **is too high** carefully. Cell clumps are not washed properly!
– The samples were not kept on ice during the last steps. It is essential that the samples be kept cold. Otherwise the staining background increases.

– BrdU is labile and sensitive to light. It is important that the BrdU **Cells are** solution is kept dark. Prepare fresh solution and repeat the labeling. **not labeled**
– Were second staining steps performed? Perhaps you left out one of the **by BrdU** many steps needed?

– After fixation the speed for centrifugation must be increased; otherwise **No cells left** the cells do not sediment. **in the tube**

– The antibodies for cell surface staining must be tested in the fixation **Cell surface** procedure. Not all antibodies survive this fixation. **staining is lost**
– Only bright stainings can be used as the staining intensity drops after fixation. Use a different antibody. Example: The murine CD45R (B220) antigen can be labeled by many different antibodies. The staining of B cells by the most commonly used antibody RA3-6B2 does not survive the fixation procedure. The antibody RA3.A1.CL6 works fine.

References

1. Houck, D.W. and Loken, M.R.: Simultaneous analysis of cell surface antigens, bromodeoxyuridine incorporation and DNA content. Cytometry 6, 531 (1985).
2. Porstmann, T., Ternynck, T. and Avrameas, S.: J. Immunol. Meth. 82: 169.(1985).
3. Gonchoroff, N.J., Katzmann, J.A., Currie, R.M.m Evans, E.L., Houck, D.W., Kline, B.C., Greipp. P.R. and Loken, M.R.: J. Immunol. Meth., 97. (1986).
4. Förster, I., Vieira, P. and Rajewsky, K.: Flow cytometric analysis of cell proliferation dynamics in the B cell compartment of the mouse. Internat. Immunol. 1, 321-331 (1989).

5. Förster, I. and Rajewsky, K.: The bulk of the peripheral B-cell pool in mice is stable and not rapidly renewed from the bone marrow. Proc. Natl. Acad. Sci. U.S.A., 87, 4781-4784 (1990).
6. Schittek, B. and Rajewsky, K.: Maintenance of B-cell memory by long-lived cells generated from proliferating precursors. Nature, 346, 749-751 (1990).
7. Schittek, B., Rajewsky, K. and Förster, I.: Dividing cells in the bone marrow and spleen incorporate bromodeoxyuridine with high efficiency. Eur. J. Immunol. 21, 235-238 (1991).

9 High-Resolution Cell Cycle Analysis: The Flow Cytometric Bromodeoxyuridine-Hoechst Quenching Technique

M. KUBBIES

9.1 Background

Cell cycle analysis of in vitro cell cultures is of relevance in basic and clinical research in various fields of immunology, cell biology and oncology. Historically cell proliferation has been studied using cell-counting techniques or radioactive thymidine labeling of S phase cells. With the advent of single cell analysis via flow cytometry interphase G_1, S, and G_2M cells were discernible. In addition, immunocytochemical techniques have been introduced selectively to label cycling $G_1/S/G_2M$ phase cells (proliferation markers such as like proliferating cell nuclear antigen (PCNA) monoclonal antibodies, mAbs) or proliferating, BrdU-labeled S phase cells (BrdU mAbs)[1].

However, none of biochemical or cytometric techniques mentioned above is able to resolve the complete history of cell proliferation of synchronous (e.g., initially resting lymphocytes) or asynchronous cell populations (e.g., tumor cell lines) or the heterogeneity of the proliferative status of cell populations. Even in clonally derived cell populations the duration of the cell cycle in different cells is heterogenous, and normal diploid cell fractions always exhibit a fraction of noncycling G_0/G_1 cells. To reveal the complexity of the proliferative status, cell cycle distribution, and kinetics of in vitro cell cultures, continuous labeling of cells is required. The most common labeling technique is the substitution of thymidine in DNA by BrdU. In contrast to the BrdU pulse labeling, BrdU mAb technique, however, the flow cytometric BrdU-Hoechst quenching analysis is a continuous BrdU labeling procedure [2–4]).

The cells are cultivated in the presence of BrdU, and during the observation period the DNA becomes unifilarily labeled in the first cell-cycle, and uni-and bifilarily in the second, third, and subsequent cell cycles [5]. Theoretically the maximum rate of the substitution of thymidine by BrdU corresponds to 50%, 75%, 87.5%, and so on for the first, second, third, and subsequent cell cycles, respectively. After BrdU labeling the cells are permeabilized and stained with the DNA-specific fluorochromes Hoechst 33258 or 33342. In the presence of BrdU the fluorescence intensity of DNA-bound Hoechst fluorochromes is decreased significantly (quenching effect) [6]. The more the cells are labeled with BrdU the more the Hoechst fluorescence intensity decreases. To achieve a better resolution of the different cell cycles, cells are counterstained with BrdU-non-sensitive fluorochromes ethidium bromide (EB) or propidium iodide (PI). The fluorochrome-labeled cells are excited with UV light, and the analysis of the emitted blue (Hoechst) and red fluorescence (EB or PI) via flow cytometry routinely reveals up to three subsequent cell cycles (sometimes the fourth cell cycle is recognized) [3].

Due to this high-resolution cell kinetic analysis of cells in the first, second, and third cell cycles the dilution effect of non- or slowly cycling cells by rapidly proliferating populations can be calculated. For the first time, the BrdU-Hoechst quenching technique enables the quantitation of the true numbers of cells of the initial population remaining either in the noncycling G_0/G_1,compartment or the cycling/dividing cells in the various cell cycles [3]. In addition, harvests and BrdU-Hoechst analysis of cells at short intervals after culture setup enables quantitative exit kinetic/cell cycle analysis of cells from all cell cycles/compartments: percentages of noncycling cells, mean and shortest cell cycle/compartment durations (Smith and Martin exit kinetic model) [3,4,7].

The BrdU-Hoechst quenching technique is applicable to initially synchronous or asnychronous cell cultures [8]. It has been used for cell kinetic analysis of numerous normal diploid cell systems [e.g., peripheral blood lymphocytes (PBLs; T and B), fibroblasts, amniotic fluid cells] and various permanent cell lines (e.g., HL60, MOLT3, Jurkat, NIH3T3, MPCll plasmacytoma, cytotoxic T Iymphocyte line (CTLL)) of human and murine origin, and other species. Due to the low BrdU substitution rate in vivo and due to the detection limit of the quenching effect, however, this analysis is not useful for in vivo BrdU labeling procedures. Finally, the BrdU-Hoechst quenching technique is a mild DNA staining procedure which enables unlimited use of immunocytochemical labeling of other intracellular or surface membrane epitops of interest [9]. In addition, due to the simplicity of cytochemistry there is only little loss of cells during the DNA staining procedure.

9.2 Material

– Synchronous, asynchronous cell cultures (adherent, suspension).
– Flow cytometer with UV excitation capability (BrdU-Hoechst only). Optional dual laser UV and 488-nm excitation for subpopulation specific BrdU-Hoechst analysis. No special requirements for data acquisition. Direct cytogram storage and data transfer capability to PC/AT computers is preferred. Software for data analysis should enable framing of regions of interest of cytograms, rotation, x- and y-axis projection, and storage of projected histograms.

Filters – **Optical filters (laser based flow cytometer):**
UV scatter light is selected by a beamsplitter DC 375 longpass (Omega) and UG11 bandpass (Schott); the 488-nm scatter light by a neutral beamsplitter 8/90 (Oriel) and a 488/10-nm bandpass (Omega).

- **Emission filters (DNA only, single laser):**
  Blue (Hoechst): K 45 bandpass (Balzers)
  Red [PI, EB, 7-aminoactinomycin D (7-AAD), LDS 751]: RG 630
  longpass (Schott)
  Beamsplitter blue/red: DC 560 longpass (Omega)
- **Emission filters [DNA and fluorescein (FITC), dual laser]:**
  Blue (Hoechst): K 45 bandpass (Balzers)
  Red (PI, EB, 7-AAD, LDS 751): RG 630 longpass (Schott)
  Green (FITC): DF 530/30 bandpass (Omega)
  Beamsplitter #1 green-blue: DR 510 longpass (Omega)
  Beamsplitter #2 green-red: DR 595 longpass (Omega)
- **Emission filters [DNA and FITC/phycoerythrin (PE), dual laser]:**
  Blue (Hoechst): K 45 bandpass (Balzers)
  Red (7-AAD, LDS 751): RG 630 longpass (Schott)
  Green (FITC): DF 530/30 bandpass (Omega)
  Orange (PE) DF 580/20 bandpass (Omega)
  Beamsplitter #1 1st/2nd laser: half mirror
  Beamsplitter #2 green-orange: DR 560 shortpass (Omega)
  Beamsplitter #3 blue-red: DC 650 longpass (Omega)
- **Optical filters (arc lamp based flow cytometer):**
- Excitation filters:
  KG1, BG 38, UG1, beamsplitter LP 450 longpass (Schott)
- **Emission filters:**
  Blue (Hoechst): K 45 bandpass
  Red (PI, EB, 7-AAD): full mirror (Schott), RG 630 longpass
  Beamsplitter blue/red: beamsplitter LP 510 longpass (Schott)

- Bovine serum albumin (BSA; Boehringer Mannheim) - RNAse A (Boehringer Mannheim) - Trizma base (Sigma) - Sodium chloride (Merck) - Magnesium chloride (Merck) - Calcium chloride (Merck) - Non-idet P40 (NP40; Boehringer Mannheim) - Hoechst 33258 (Boehringer Mannheim) - PI (Sigma) - EB (Boehringer Mannheim) - 7-AAD (Serva) - LDS 751 (Exciton)	**DNA staining**
- BrdU (Boehringer Mannheim) - Deoxycytidine (DC; Sigma) - Dimethylsulfoxide (DMSO; Sigma) - Cell culture media (Boehringer Mannheim) - Fetal calf serum (FCS; Boehringer Mannheim) - 25 and 75 cm^2 cell culture flasks (Falcon) 15-ml centrifuge tubes (Falcon)	**Cell culture, cell harvest**

Miscellane- – Minifuge (Heraeus)
ous – MPLUS software for data analysis (Phoenix Flow Systems)

9.3 Method

Cell culture
1. Cultivate cells at cell densities of 1–5x10^5 cells/ml (suspension culture) or 1–5x10^5 cells/25 cm^2 flask using routine cell culture techniques.
2. Supplement medium with 8x10^{-5} M BrdU and DC and protect cells from light during cultivation.
3. Harvest cells at room temperature by centrifugation at 400 g for 10 min (avoid direct exposure by intense light; low light level acceptable).
4. Cells may be stained directly for flow cytometric analysis. For short- or long-term storage cells are frozen in freezing medium at -20^{0}C (basal culture medium, 10% FCS, 10% DMSO).

DNA staining Fluorochrome stock solutions (1 mg/ml) are prepared in sterile aqua dest (PI, EB), methanol (LDS 751) or 10% DMSO (7-AAD), protected from direct light exposure and stored at 4°C. Stock solutions may be used up to 1 year. Inactivate residual DNAse activity in RNAse A by heating of RNAse stock solution for 2 h at 80°C or use DNAse free RNAse preparations.
1. Thaw frozen cells or use freshly isolated cells, and centrifuge at room temperature at 400 g for 10 min.
2. Resuspend cell pellet at 1–5x10^5 cells/ml in DNA-staining buffer (100 mM Tris 7.4, 154 mM NaCl, 1 mM CaCl$_2$, 0.5 mM MgCl$_2$, 0.2% BSA, 0.1% NP40) supplemented with 10 U/ml RNAse A and 1.2 µg/ml Hoechst 33258.
3. Incubate at 4°C for 15 min in the dark.
4. Add BrdU-nonsensitive fluorochromes to final concentrations of: 2 µg/ml PI or EB, 10 µg/ml 7-AAD, or 40 µg/ml LDS 751 (use 7-AAD or LDS 751 only with FITC- and PE-labeled cells).
5. Incubate cells at least for a further 15 min at 4°C in the dark, and analyze sample within 8 h.

Subpopu-
lation specific
DNA staining
1. Use freshly isolated cells and label cells with mAbs via direct or indirect FITC and/or PE immunofluorescence.
2. Resuspend pellet in 50 µl PBS vortex cells and add 1 ml 1% paraformaldehyde (4°C).
3. Fix cells for 30–60 min at 4°C.
4. Centrifuge cells at 400 g for 10 min at 4°C and resuspend pellet at 1–5x10^5 cells/ml in PBS/0.01% Tween 20.

5. Add Hoechst 33258 (1.2 µg/ml; final concentration) and proceed according DNA staining step 3. Use PI or EB if cells are FITC labeled only. Use 7-AAD or LDS 751 with PE-labeled cells.
6. Final incubation time prior flow cytometric analysis should be at least 2h.

1. Select UV light in arc lamp based flow cytometers by using the excitation filters as indicated in Sect. 9.2 (HBO 100-W/2 arc lamp). In laser-based instruments use either UV-capable ion argon lasers (tune to 351/364 nm and 50 mW) or HeCd lasers (325 nm and 10 mW miniumum) for excitation. Gate on DNA or scatter signal. **Flow cytometric analysis**
   For subpopulation-specific BrdU-Hoechst analysis in laser-based instruments use an argon laser additionally tuned at 488 nm and 50 mW for FITC and PE excitation. Delay both lasers for 30–60 µs. First/second laser adjustment depends on optical configuration of instrument. Gate on DNA or scatter signal.
2. Run samples at 300–1000 cells/s at room temperature (wait 30 s before data recording).
3. Record blue Hoechst fluorescence and red PI, EB, 7-AAD, or LDS 751 fluorescence using the optical filters indicated in the Sect. 9.2.
4. Depending on the complexity of the bivariate BrdU-Hoechst PI cytogram (or EB, 7-AAD, or LDS 751) analyze at least 2×10^4 cells. Analyze at least 5×10^4 cells in subpopulation-specific BrdU-Hoechst analysis.
5. Store data preferentially as single cytograms. In subpopulation-specific analysis store data in list mode for subsequent gating procedures.

1. Quantitate cell cycle distributions by extracting individual first, second, and third cell cycles from the bivariate BrdU-Hoechst PI cytograms, and perform G_1-S-G_2M DNA histogram analysis of individual cell cycles (e.g., use MPLUS software for both procedures; $G_0/G_1 + S + G_2M = 100\%$). **Data analysis**
2. Quantitation of real data (represents cell culture status): Calculate the relative percentages of cells in the first, second, and third cell cycles, and multiply by the percentages of the G_1, S and G_2M compartments of individual cell cycles from step 1. This gives the percentages of cells in individual compartments of the various cell cycles.
3. Quantitation of original data (representing true numbers of cycling or non-dividing cells; dilution effect of rapidly proliferating cells exerted on slowly or nonproliferating populations is taken into account): Take data from step 2; calculate „dilution factor" (DF):

$$DF = 100 / [(G_0G_1 + S + G_2M) + (G_1' + S' + G_2M')/2 + (G_1'' + S'' + G_2M'')/4].$$

In the case of a fourth cell cycle the percentages of real data are divided by 8 and added to the denominator as shown above.

first cycle: % orig. G_0/G_1 = % real G_0/G_1 x DF

% orig. S   = % real S   x DF

% orig. G_2M = % real G_2M x DF

second cycle: % orig. G_1'   = % real G_1'/2   x DF

% orig. S'   = % real S'/2   x DF

% orig. G_2M' = % real G_2M'/2 x DF

third cycle: % orig. G_1"   = % real G_1"/4   x DF

% orig. S"   = % real S"/4   x DF

% orig. G_2M" = % real G_2M"/4 x DF

The „percentage increased cell number" corresponds to:
% inc. cells = DF x 100
The calculation of the „Smith and Martin exit kinetic curves" of the individual cell cycle compartments requires specialized computer software. The formulas for calculation of the minimum and mean duration of cell cycle compartments, and the fractions of noncycling cells are shown in [3] and [10].

9.4 Example

The complexity of the cell kinetics of human PBLs is revealed only by the BrdU-Hoechst quenching technique. Typical examples of 48 and 72 h harvests of phytohemagglutinin stimulated, human PBLs are shown in Figure 1 (panels A and B). Two days after polyclonal activation the majority of PBLs are still in the noncycling G_0/G_1 compartment, and some cycling cells are found in the first (S, G_2M) and second cycle (G_1', S', G_2M'). The power of the flow cytometric BrdU quenching technique is shown by the fact that even the very small fraction 0.3% of third cycle G_1" phase cells is significantly resolved. The nuclear decay lane represents dead cells of the G_0/G_1 population.

Three days after stimulation most of the PBLs have left the noncycling G_0/G_1 compartment. The culture now consists of cells in the first, second and third (G_1", S", G_2M") cell cycles. The fact that there are still PBLs in the S phase of the first cell cycle indicates that the process of cell activation and recruitment is still going on in the G_0/G_1 population. To display the quantitative distribution of the PBLs the isometric plots of panels A and B are shown below the bivariate BrdU-Hoechst PI dot plots. The corresponding framed regions of the individual first, second and third cell cycles in panels A and B are shown as one-parametric DNA-histograms.

Table 1 displays the dilution effect of fast proliferating, dividing cells (normally overestimated) exerted on slow cycling or nonproliferating cell populations (normally underestimated). Data are calculated from Figure 1

Example   81

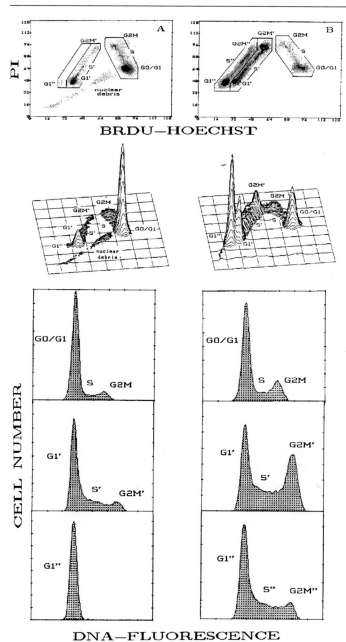

Figure 1. Bivariate BrdU-Hoechst Pl cytograms of phytohemagglutinin-stimulated PBLs harvested 48 (A) and 72 h (B) after activation. Due to the quenching effect of BrdU exerted on the DNA-bound Hoechst fluorochromes the lanes of subsequent cell cycles *(framed areas)* are left to the previous one. From the top to the bottom the data are displayed as bivariate dot plot, isometric plot, and projected first, second, and third cell cycle DNA histograms.

First cell cycle: G_0/G_1, S, G_2M.
Second cell cycle: G_1', S', G_2M'.
Third cell cycle: G_1'', S'', G_2M''.

panel A (72 h harvest). With the knowledge of the number of cells in different, subsequent cell cycles the true percentages of cycling and nonactivated cells can be calculated.

A PBL subpopulation specific, multiparameter BrdU-Hoechst analysis is shown in Figure 2. PBLs from patients with Fanconi's anemia display a significant slowing of cell proliferation due to a slowing of G_2 phase progression and G_2 phase arrest in the first and second cell cycle. However, cell activation of G_0/G_1 cells is not affected [11]. From panels A and B it is obvious that almost all CD8+ T cells have left the G_0/G_1 compartment. A significant number of G_0/G_1 cells are present only in the CD4-/CD8- population (mainly B cells, panel B). The G_2 phase arrested populations of CD8+ and CD8- cells are indicated by arrows.

Table 1. Calculation of the real data (cell culture status) and original data (true percentages of cycling/noncycling cells of the original population of the cell cycle distribution from the accompanying figure).

	First cell cycle			Second cell cycle			Third cell cycle		
	G_0/G_1	S	G_2M	G_1'	S'	G_2M'	G_1"	S"	G_2M"
Real (%)	14.3	4.8	3.3	8.3	20.8	6.7	18.5	21.5	1.8
Original (%)	28.2	9.4	6.4	8.2	20.6	6.6	9.0	10.6	0.9

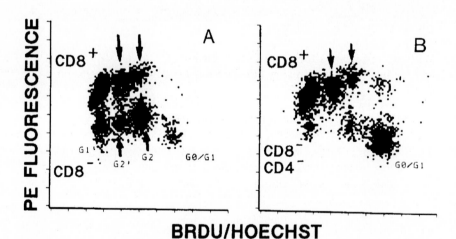

BRDU/HOECHST

Figure 2. T cell subpopulation specific, high-resolution cell kinetic BrdU-Hoechst analysis of Fanconi's anemia cells 72 h after phytohemagglutinin-activation. Gating of CD4-FITC and CD8-PE labeled, and BrdU/Hoechst stained PBLs is on the CD8+ and CD8- fractions (includes CD4+ T cells; A) and on the CD8+ and CD4-/CD8- populations (B)

9.5 Modifications

Instead of detergent lysis of cells for DNA staining methanol (90%) or ethanol (70%) fixation procedures can be used for the bivariate BrdU-Hoechst PI cell cycle analysis. The latter fixation techniques additionally enable multiparameter analysis of intracellular epitops. Cells should be fixed for at least 1 h at -20°C prior to subsequent immunofluorescence labeling (depending on the antibody permeabilization by 0.05%–0.1% NP40/PBS, Tween 20/PBS, or Triton X100/PBS may be required). DNA staining is performed thereafter as described above for the paraformaldehyde fixation technique used for surface marker labeling.

9.6 Tips, Tricks, and Troubleshooting

The BrdU-Hoechst quenching is dependent on a variety of cell biological, biochemical, biophysical, and technical factors [3]. For example, it is species (AT/GC ratio) and cell type dependent (intracellular thymidine pool), and might even depend on the serum lot used for cell culture (thymidine content). In order to block thymidine incorporation into DNA and therefore increase the amount of BrdU in DNA, fluorodeoxyuridine can be used as inhibitior for deoxythymidylate synthetase. However, technical and cellular parameters should be optimized first instead of using a potentially toxic inhibitor. **Fluorodeoxyuridine**

Without significant alteration of the quenching effect the Hoechst and EB or PI concentrations can vary between 1.0–1.5 and 1.5–2.5 µg/ml, respectively. **Hoechst, EB, and PI concentration range**

Depending on the cell type and optical instrument configuration the range of the BrdU concentration for optimal quenching is $2-10 \times 10^{-5} M$. At higher concentrations inhibition of cell activation and cell cycle progression may (but does not necessarily) occur. Equimolar concentrations of deoxycytidine should be added as medium supplemented to avoid possible adverse proliferative effects due to disturbance of the intracellular nucleotide pool. **BrdU concentration**

Cell densities significantly above 5×10^5 cell/ml in the staining buffer can decrease the quenching effect. **Cell concentration**

Alcohol- or paraformaldehyde-fixed DNA-stained cells can be stored for several days with loss of resolution of the bivariate BrdU-Hoechst PI cytogram. **Fixed cells**

7-AAD and LDS 751 are recommended as BrdU nonsensitive counterstains using PE-labeled cells. **7-AAD and LDS 751**

HeCd lasers The first cell cycle G_2M/G_1 fluorescence ratio is decreased below 2.0 due to overlapping effects of red fluorochrome (e.g. PI) excitation via the UV light source and from energy transfer from the Hoechst fluorochromes. This phenomenon is lacking with 325-nm excitation from a HeCd laser.

Filter selection The optical filter for blue (Hoechst) and red (PI, EB, etc.) fluorescence analysis in Sect. 9.2 are recommodations only. The blue and red fluorescence should be recorded at 400–500 nm and above 580 nm, respectively. The 50% value of the dichroic beamsplitter should be in the range between 510–570 nm.

Aggregates All DNA staining procedures induce cell clumping and increase the G_2M phase fractions artificially (depending on cell type, e.g., adherent cells). This is more significant using alcohol fixation procedures. To avoid cell clumping the „peak versus area" gating procedure is recommended for the sensitive, high-resolution cell kinetic BrdU-Hoechst technique.

Acknowledgement. The Fanconis anemia cells were kindly provided by Dr. D. Schindler from the Department of Human Genetics University of Wurzburg, FRG.

References

1. Gray JW Darzynkiewcz Z, (eds.): Techniques in cell cycle analysis. Humana Press, Clifton, NJ, 1987.
2. Kubbies M, Rabinovitch PS: Flow cytometric analysis of factors which influence the BrdU-Hoechst quenching effect in cultivated human fibroblasts and lymphocytes. Cytometry 3:276,1983.
3. Kubbies M, Hoehn H, Schindler D, Chen YC, Rabinovitch PS: Cell cycle analysis via BrdU-Hoechst flow cytometry: principles and applications. In: Flow Cytometry (Volume ll), Yen A (ed), CRC- Press, Florida, p. 5,1989.
4. Poot M, Kubbies M, Hoehn H, Grossmann A, Chen Y, Rabinovitch PS: In vitro cell cycle analysis using continuous BrdU labeling and bivariate Hoechst 33258-EB flow cytometry. In: Methods in cell biology, 33, Darzynkiewcz Z, et al (eds); Academic Press, NY, pp. 186,1990.
5. Kubbies M, Friedl R: Flow cytometric correlation between BrdU-Hoechst quench effect and base pair composition in mammalian cell nuclei. Histochemistry 83:133, 1985.
6. Latt SA, Wohlleb JC: Optical studies of the interaction of 33258 Hoechst with DNA, chromatin and the interphase chromosome. Chromosoma 52:297, 1975.
7. Rabinovitch PS: Regulation of human fibroblast growth rate by both non-cycling cell fraction and transition probability is shown by growth in BrdU followed by Hoechst 33258 flow cytometry. PNAS 80:2951,1983.
8. Omerod MG, Kubbies M: Cell cycle analysis of asynchronous cell populations by flow cytometry using bromodeoxyuridine label and Hoechst-propidium iodide staining. Cytometry, in press, 1992.
9. Kubbies M: High resolution multiparameter cell cycle analysis by BrdU/ Hoechst flow cytometry. In: Progress in cytometry: flow and image. Reports from the 3rd European cytometry users conference 1989 in Ghent, Belgium. Becton Dickinson (eds), p. 30,1990.

10. Kubbies M: Alteration of cell cycle kinetics by reducing agents in human PBL's from adult and aged donors. Cell Proliferation, 25:157,1992.
11. Kubbies M, Schindler D, Hoehn H, Schinzel A, Rabinovitch PS: Endogenous blockage and delay of the chromosome cycle despite normal recruitment and growth phase explain poor proliferation and frequent endomitosis in Fanconi anemia cells. Am J Hum Genet 37: 1007,1985.

Part IV Cellular Activation and Biochemistry

10 Cell Activation: Indo-1 Ratiometric Analysis of Intracellular Ionized Calcium

M. KUBBIES

10.1 Background

Intracellular ionized calcium (Ca^{2+}) regulates various metabolic processes and is involved in signal transduction and cell activation. The intracellular concentration of Ca^{2+} in resting cells (100–200 nM) is far below the concentration of the extracellular environment. Most of the intracellular calcium, however, is bound in its nonionized form in the endoplasmic reticulum, mitochondria, cytosol, and cell memebrane. Influx of Ca^{2+} into cells occurs through the action of voltage gated channels after membrane depolarization or by the action of receptor-gated channels. The excess of Ca^{2+} is pumped out of the cells by the action of a membrane-bound Ca^{2+} ATPase (for review see [1,2]).

An increase in intracellular Ca^{2+} by extracellular binding of agonists to receptors initiates the activation cascade of the phosphoinositol pathway [3,4]. After binding to the receptor phospholipase C is activated, which cleaves phosphatidylinositol-4,5-bisphosphate into both second messengers diacylglycerol and inositol-1,4,5-trisphosphate. The latter binds to receptors of the endoplasmic reticulum, and as a first step Ca^{2+} is released from this intracellular compartment. Subsequently Ca^{2+} influx occurs from the extracellular environment. Although it remains open to debate whether intra- and/or extracellular influx of Ca^{2+} is sufficient for cell activation, a subsequent cascade of still unknown events triggers the cell activation process toward proliferation of the cells. However, there is evidence that cell proliferation can occur in the absence of a continuous increase in intracellular Ca^{2+} [5,6]. It remains to be shown whether small local changes or oscillations in intracellular Ca^{2+} are sufficient for cell activation [7].

Historically, intracellular Ca^{2+} was analyzed by microelectrode techniques and loading of cells by $^{45}Ca^{2+}$. Since then, various Ca^{2+}-binding fluorochromes have been developed, changing either the fluorescence intensity (quin-2, fluo-3), absorption spectrum (fura-2), or emission spectrum (indo-1) upon binding of ionized calcium (for review see [1]). Analysis of the Ca^{2+}-dependent alteration in fluorescence intensity requires simpler technical flow cytometric equipment. Recently fluo-3 has been developed to enable excitation at 488 nm. However, inhomogeneity of dye loading and different cell sizes may result in a rather high coefficient of variation (CV) of the fluo-3 histogram [8]. These limitations are circumvented using ratio analysis of fluorescence signals of cells loaded with Ca^{2+}-sensitve fluorochromes. Ratio analysis yields results independent of biological (cell size), cytochemical (dye loading), or technical variations (drift in excitation light intensity). Whereas alteration in the absorption spectrum is ideally suited for microscopic digital ratio imaging (dual wavelength excitation, single wavelength emission analysis, fura-2), changes in the emission spectrum are optimal for flow cytometric analysis (single wavelenght excitation, dual wavelenght emission analysis, indo-1) [8-10].

Indo-1 exhibits a Ca^{2+}-independent emission of fluorescence intensity at the isobestic point around 450 nm (355 nm excitation). Binding of Ca^{2+} to indo-1 decreases the blue fluorescence intensity above 450 nm, whereas the intensity of the violet fluorescence below 450 nm increases [11]. Therefore, the ratio of violet/blue fluorescence increases even more compared to the increase in violet fluorescence intensity only. It has been shown that the indo-1 ratio varies with the emission wavelength recorded. Maximum ratios are obtained collecting light around or below 400 nm (violet) and above 500 nm (blue) [9]. Additionally, due to ratio imaging, the CV of the ratio distribution of the violet and blue fluorescence of a nonactivated resting cell population is significantly smaller. This improvement in flow cytomtric resolution enables detection of even small releases of intracellular Ca^{2+} from endoplasmic stores after buffering of the extracellular Ca^{2+} by ethyleneglycoltetraacetic acid (EGTA) [9,12].

The calcium-sensitive fluorochromes are membrane impermeable. Loading of viable cells is performed using indo-1 prepared as acetoxymethyl (AM) esters. These AM esters become hydrolyzed by intracellular esterases, and indo-1 is trapped within the cells (5-20 μM). Dead cells do not accumulate calcium sensitve fluorochromes and are easily gated out during flow cytometric analysis. The concentrations of indo-1 used for flow cytometric analysis have been shown to be nontoxic to peripheral blood lymphocytes (PBLs) [9].

The requirement of UV excitation for indo-1 is regarded as a limitation for its widespread use in flow cytometry. However. using a long-lived inexpensive HeCd laser for indo-1 excitation gives excellent resolution of analysis of intra- and extracellular Ca^{2+} influx [13]. In addition. the use of dual lasers with indo-1 enables subpopulation-specific analysis in heterogenous cell systems without fluorescence overlap of fluorescein isothiocyanate (FITC) and/or phycoerythrin (PE) used as surface marker labels of monoclonal antibodies (mAbs) [8-10,12].

10.2 Material

– Synchronous or asynchronous suspension cell cultures or freshly isolated cells sensitive to receptor stimulation.
– Flow cytometer with UV excitation capability (indo-1 only). Optional dual laser UV and 488-nm excitation for subpopulation (FITC and/or

PE) specific indo-1/Ca^{2+}analysis. Gating of violet and blue indo-1 fluorescence signals and signal rationing are required for data aquisition. Direct cytogram storage and data transfer capability to PC/AT computers is preferred. Software for data analysis should enable the calculation of both the percentage Ca^{2+} positive cells and the indo-1 ratio of Ca^{2+} positive cells versus resting population as a function of time and display of these data as time histograms.

Filters
– **Optical filters:** UV right angle scatter is selected by a beamsplitter DC 375 longpass (Omega) and UG 11 bandpass (Schott). The UV forward scatter signal (UG 11 bandpass) is used as trigger signal.
– **Emission filters (indo-1 only, single laser):**
Violet: DF 395/25 bandpass (Omega)
Blue: DF 510/30 bandpass (Omega)
Beamsplitter violet/blue: DC 420 longpass (Omega)
– **Emission filters (indo-1 and PE, dual laser):**
Violet: DF 395/25 bandpass (Omega)
Blue: DF 510/30 bandpass (Omega)
Orange (PE): DF 580/20 bandpass (Omega)
Beamsplitter #1 violet/blue: DC 420 longpass (Omega)
Beamsplitter #2 orange/blue: DC 560 shortpass (Omega)
– **Emission filters (DNA and FITC/PE, dual laser):**
Violet: DF 395/25 bandpass (Omega)
Blue: DF 510/30 bandpass (Omega)
Green (FITC) DF 530/30 bandpass (Omega)
Orange (PE): DF 580/20 bandpass (Omega)
Beamsplitter #1 1st/2nd laser: half mirror
Beamsplitter #2 green-orange: DR 560 shortpass (Omega)
Beamsplitter #3 violet/blue: DC 420 longpass (Omega)

Indo-1 staining and cell activation
– Indo-1 AM ester (Boehringer Mannheim)
– Water free dimethylsulfoxide (DMSO; Sigma)
– Thrombin (Sigma) formyl-methionyl-leucyl-phenylalanine (fMLP; Sigma)
– Mouse CD3 (Boehringer Mannheim)
– Goat „mouse" (Boehringer Mannheim)
– Synperonic F127 (Boehringer Mannheim)
– Ionomycin (Calbiochem)
– EGTA (Merck)

Cell culture
– Cell culture media (Boehringer Mannheim)
– Fetal calf serum (FCS; Boehringer Mannheim)
– 15-ml centrifuge tubes (Falcon)

Miscellaneous
– Minifuge (Heraeus)
– Fluorescence microscope
– MTIME software for data analysis (Phoenix Flow Systems)

10.3 Method

Indo-1 staining Indo-1 AM stock solution (1 mg/ml) is prepared in water-free DMSO, and small aliquots are stored at -20°C in the dark. Stock solutions may be used up to several months. EGTA is dissolved in water as 100 mM stock adjusted to pH 7.4, synperonic F127 (or pluronic F127) as 20% (wt/vol) in DMSO, and ionomycin as 2 mg/ml in absolute ethanol. Avoid the use of sodium azide in the labeling medium throughout indo-1 AM loading. Use sodium azide free mAbs for labeling of cells.

1. Harvest cells by conventional techniques. Centrifuge at 200 g at room temperature for 10 min.
2. Resuspend cell pellet at 5x10^6 cells/ml in RPMI 1640/10% FCS. Add 3 µg/ml indo-1 AM, and mix gently.
3. Incubate at 37°C for 45 min in the dark.
4. Centrifuge cells at 200 g for 10 min, remove supernatant completely, and resuspend pellet carefully in RPMI 1640/10% FCS at 1x10^6 cells/ ml.
5. Keep samples at room temperature in the dark and raise to 37°C prior to analysis. Analyze samples within 2–3 h. Control indo-1 loading of cells by microscopic analysis. Cells should be labeled brightly and uniformly, showing no intracellular bright spots of Indo-1 fluorescence (compartmentation of indo-1)

Subpopulation-specific indo-1 staining
1. Indo-1 loading as described above (steps 1–4) except resuspension of cells at 1x10^6 cells/100 µl RPMI 1640/10% FCS.
2. Label cells not of interest for Ca^{2+} analysis using a cocktail of FITC- and/or PE-labeled mAbs for 20 min at room temperature in the dark. Avoid labeling of cell population of interest due to possible interference of the receptor binding of mAb with Ca^{2+} signal activation (negative gating procedure; see below).
3. Resuspend cells in 2 ml RPMI 1640/10% FCS.
4. Centrifuge cells at 200 g for 10 min. Remove supernatant completely and resuspend pellet carefully in RPMI 1640/10% FCS at 1x10^6 cells/ ml.
5. Keep samples at room temperature in the dark and raise to 37°C prior analysis. Analyze samples within 2–3 h.

Flow cytometric analysis
1. Tune argon laser to 351/364 nm and 50 mW or use HeCd laser at 325 nm and 10 to 30 mW for excitation. Select viable cells by gating on forward scatter signals. For subpopulation specific indo-1 analysis use a second argon laser tuned at 488 nm and 50 mW for FITC and/or PE excitation. Delay both lasers for 30–60 µs. First/second laser adjustment depends on optical configuration of instrument.
2. Run samples at 200–600 cells/s at 37°C. Use saline or phoshate-buffered saline as sheath fluid.
3. Record blue and violet fluorescence as linear signals (exclude lowest and highest channels from ratio analysis). Adjust maximum of blue and violet histograms of resting cells in the upper and lower half of the histogram, respectively. Gate on scatter characteristics of cells (e.g.,

Example   93

lympho-, mono-, or granulocytes; exclude dead cells from analysis) and on non-mAb-labeled cell fraction, performing subpopulation specific Ca^{2+} analysis (negative gating).

4. Analyse nonactivated cells and adjust gaussian distributed violet/blue fluorescence ratio signals in the lower region of the ratio histogram.

5. Stop sample flow, add activating agonist to cells, and restart analysis as quickly as possible.

6. Store indo-1 violet/blue ratio histograms as a function of time (time slices) preferentially as cytograms. In subpopulation-specific analysis obtain several indo-1 ratio/time cytograms using gating procedures.

7. Optimal indo-1 loading of cells and adjustment of instrument should be controlled by induction of a maximal Ca^{2+} response by the Ca^{2+} ionophore ionomycin (2 µg/ml). The increase in the indo-1 ratio of all human PBLs should be at about factor 6–8 over the baseline of the resting population.

8. The release of intracellular Ca^{2+} from endoplasmic reticulum is analyzed by resuspension of the cells either in Ca^{2+}-free medium or the addition of 5 mM EGTA 2–5 min prior to cell activation.

– Define region of resting and responding cell population on time axis.   **Data analysis**

– „**Percentage Ca^{2+} positive cells**": Calculate either the percentage of Ca^{2+} positive cells above an indo-1 ratio channel threshold defined by the resting population (accept 2%–5 % background of low-level Ca^{2+} positive cells; percentage above threshold) or substract resting population from the distribution of activated cells (percentage of responding cells). Calculation of the percentage of responding cells is less sensitive to deviations in the indo-1 ratio histogram from a normal distribution [14].

– „**Increase in indo-1 ratio**": Calculate the mean indo-1 ratio of the Ca^{2+} responding cell population versus the resting nonactivated cell fraction. The intracellular concentration of ionized Ca^{2+} is calculated using the formula and calibration procedure described by Grynkiewcz and Rabinovitch [9, 11].

– Calculation of the percentage of Ca^{2+} positive cells and the absolute and relative values of intracellular Ca^{2+} is implemented in the kinetic analysis program MTIME (written by P.S. Rabinovitch).

10.4 Example

Figure 1 illustrates the Ca^{2+} response of human PBLs activated by incubation with CD3 followed by <mouse> cross-linking (panel A) and human thrombocytes stimulated with thrombin (panel B). Human PBLs and thrombocytes were isolated according to standard procedures and cells were identified by their forward/right-angle scatter characteristics. The resting populations of both cell types exhibit a gaussian distributed shape

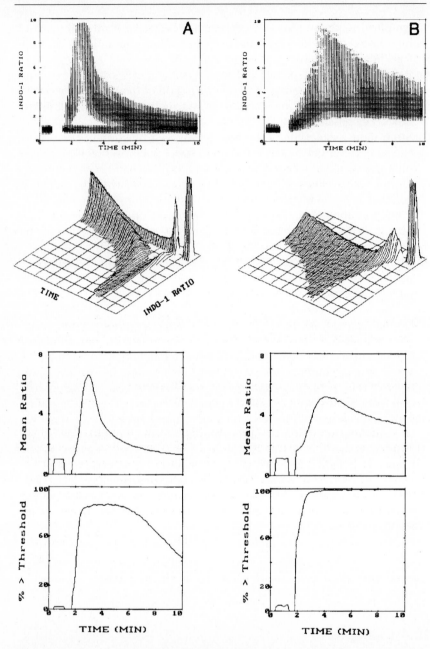

Figure 1. Increase in intracellular Ca^{2+} in human lymphocytes (A) and thrombocytes (B) activated with <CD3>/<mouse> (5 µg/ml each), and thrombin (1 µg/ml), respectively. The agonists were added 1 min after start of the experiment (see gaps). The isometric plots and the calculations of the mean ratio (indo-1 ratio of activated versus resting population), and the percentage of cells above the threshold (percentage of Ca^{2+} positive cells) are shown below to illustrate the quantitative changes in intracellular Ca^{2+}

Example   95

in the 1st min after the start of flow cytometric analysis (see isometric plots below panels A and B respectively). The gap indicates the stop of the sample flow in order to add the stimulating agonists. The flow cytometric indo-1 ratio analysis was restarted 30 s later.

The human PBLs show a rapid and high Ca^{2+} influx within 1.5 min. However. Ca^{2+} is pumped out of the PBLs rapidly and declines to almost normal values within 5 min after maximum response (panel A). It is evident that a distinct fraction of 17.8% of the PBLs is nonresponsive to <CD3> activation (cell population showing an indo-1 ratio of 1.0). In contrast. the Ca^{2+} response of thrombocytes is slower. and the intracellular Ca^{2+} level reaches its maximum 2.3 min after addition of thrombin (panel B). However all cells exhibit an increase in intracellular Ca^{2+}. The Ca^{2+} efflux from the cytoplasma of thrombocytes is slower. and almost all cells still show significantly increased intracellular Ca^{2+} levels within the observation period. The quantitative data on the kinetics of the increased indo-1 ratio and the percentage of positive cells are shown below the cytograms.

The Ca^{2+} response of human mono- and granulocytes is shown in Figure 2 (panels A and B, respectively). These cells were gated from the leukocyte forward/right-angle scatter cytogram (whole blood lysis). Compared to PBLs and thrombocytes the Ca^{2+} influx occurs more rapidly, and the Ca^{2+}-responding cells were almost positive at the restart of flow cytometric analysis. The Ca^{2+} efflux of f-MLP activated monocytes was complete within 4 min after maximum response (panel A) and approached to the baseline level in fMLP-stimulated granulocytes (panel B). However, within the observation period, the latter cells showed greater heterogeneity of intracellular Ca^{2+} levels (higher CV of the indo-1 ratio histogram).

The quantitative data from these two figures are summarized in Table 1. It is obvious that flow cytometric indo-1 Ca^{2+} flux analysis shows several advantages. Small CVs of the resting population, improving the detection limit of the Ca^{2+} response, correct quantitation of the heterogeneity of the Ca^{2+} response (e.g., nonresponding versus responding population), and second laser analysis (FITC and/or PE) enables subpopulation-specific analysis.

Table 1. Calculation of the percentage of calcium-positive cells, increase in intracellular Ca^{2+} (indo-1 ratio), and time delay to maximum Ca^{2+} response (min) from the figures (analysis performed using the MTIME program written by P.S. Rabinovitch).

	CD3: PBLs	Thrombin: thrombocytes	fMLP: monocytes	granulocytes
Indo-1 ratio	7.1	4.6	2.1	3.6
Percentage positivecells	82.2	99.5	81.6	96.0
Delay to maximum response	1.5	2.3	0.9	0.4

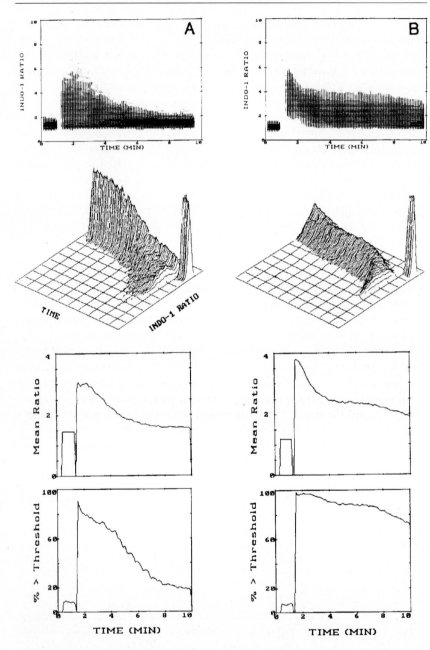

Figure 2. Intracellular release of Ca^{2+} in fMLP (1×10^{-7} *M*) stimulated human mono-cytes (A) and granulocytes (B). fMLP was added 1 min after start of flow cytometric analysls

10.5 Modifications

The concentration of cells to be loaded with indo-1 AM can vary between 1 and 5×10^6 cells/ml loading medium. Any culture medium optimal for cell viability and physiology can be used instead of RPMI. The FCS can be omitted or the concentration decreased (e.g. 1% FCS) if cells remain viable throughout the staining procedure.

Optimal loading with indo-1 may depend on cell type under investigation, and the concentration should be in the range of 1–5 µg/ml. Higher CV values of resting population and Ca^{2+} buffering can occur at lower or higher indo-l AM concentrations respectively. Loading of cells with indo-1 can be performed at lower temperatures of 25°–30°C and incubation time can be decreased to 30 min.

10.6 Tips, Tricks, and Troubleshooting

– Enrichment of indo-1 in non-Ca^{2+}-responsive intracellular compart- **Compart-** ments decreases the maximum Ca^{2+} response detectable by alteration in **mentalization** violet/blue indo-l fluorescence (control uniform loading of cells). Compartmentation can be minimized by addition of 2 µM synperonic F127 (or pluronic F127; disolve at 20% w/v in DMSO) during indo-1 AM loading of cells, and decrease of incubation temperature to 25°–30°C.

– Do not use sodium azide in the labeling medium. Use dialyzed mAb **Labeling** preparations (this may not always be required; check control sample). **medium** Cells become leaky to Ca^{2+}, and show increased indo-1 ratios. Skewing of the resting, gaussian distributed population toward higher indo-1 ratios occurs.

– Subpopulation-specific Ca^{2+} analysis requires labeling of cells by mAbs. **Controls for** However, mAb binding may induce alterations in the Ca^{2+} flux. Use **antibody** negative gating procedure: label all cells not of interest for Ca^{2+} analysis **activation** by mAbs (FITC or PE). Gate on the unlabeled cell fraction recording violet and blue indo-1 fluorescence.

– Indo-1 loading of cells can be affected by various cellular and/or **Indo-1** cytochemical factors. Some cells show no activity of intracellular **problems** esterases (loading of cells must be done by injection of indo-1). Indo-1 AM may be hydrolyzed prior use (disolve indo-1 AM in water-free DMSO and store in sealed caps at -20°C).
– Fluorescent agonists or inhibitors alter indo-1 ratios which might mimic an increase (more violet) or decrease (more blue) of the indo-1 ratio without any physiological relevance.

– Check the number of AMs coupled to indo-1 for cell loading. AMs released intracellularly by unspecific esterases may be toxic to cells. Use Indo-1 with a lower number of AM esters.

Cross-linking antibodies for activation – Although mAbs used for activation recognize identical receptors. slight differences in receptor epitope specificity or mAb affinity results in altered Ca^{2+} response. Cross-linking of mAbs by <mouse> F(ab)2 fragment may be required.

Consistency control – Analyze the Ca^{2+} flux of control cells at the end of each series of experiments to control the constancy of Ca^{2+} response.

Filter selection – Selection of optical filters for analysis of violet and blue fluorescence affects maximal response recorded. For optimal indo-1 ratio analysis use violet bandpass filters with a maximum below 410 nm. The blue fluorescence should be analyzed using longpass or bandpass filters above 490 nm.

Oscillations – Cells may show Ca^{2+} oscillations after agonist stimulation. Oscillations are not detectable by flow cytometry. Use digital video imaging for analysis of oscillatory Ca^{2+} processes.

References

1. McCormack JG, Cobbold PH (eds.): Cellular calcium: a practical approach. IRL Press, Oxford, 1991.
2. Cheung WY (ed): Calcium and cell function. Vol. Vll. Academic Press, New York, 1987.
3. Abdel-Latif AA: Calcium-mobilizing receptors, polyphosphinositides, and the generation of second messengers. Pharm Rev 38: 227,1986.
4. Linch DC, Wallace DL, O'Flynn K: Signal transduction in human T lymphocytes. Immunol Rev 95:137,1987.
5. Gelfand EW, Cheung RK, Grinstein S, Mills GB: Characterization of the role for calcium influx in mitogen-induced triggering of human T cells: identification of calcium-dependent and calcium-independent signal. Eur J Immunol 16:907, 1986.
6. Raspe E, Reuse S, Roger PP Dumont JE: Lack of correlation between the activation of the Ca^{2+} phoshatidylinositol cascade and the regulation of DNA-synthesis on the dog thyrocyte. Exp Cell Res 198:17,1992.
7. Berridge MJ, Galione A: Cytosolic calcium oscillators. FASEB 2:3074,1988.
8. Rabinovitch PS, June CH: Measurement of intracellular ionized calcium and membrane potential. In: Flow cytometry and sorting, Melamed MR, Lindmo T, Mendelsohn ML (eds). Wiley-Liss, New York, p. 651,1990.
9. Rabinovitch PS, June CH, Grossmann A, Ledbetter JA: Heterogeneity of T-cell intracellular free calcium responses after mitogen stimulation with PHA or anti-CD3: use of Indo-1 and simultaneous immunofluorescence with flow cytometry. J Immunol 137: 952,1986.
10. June CH, Rabinovitch PS: Measurement of intracellular ions by flow cytometry. In: Current protocols in Immunology. Coligan JE, et al (eds.), John Wiley & Sons, New York, p. 5.5.1.,1991.
11. Grynkiewcz G, Poenie M, Tsien R: A new generation of Ca^{2+} indicators with greatly improved fluorescence properties. J Biol Chem 260:3440,1985.

12. Kubbies M, Goller B, Russmann E, Stockinger H, Scheuer W: Complex Ca^{2+} flux inhibition as primary mechanism of staurosporine induced impairment of T-cell activation. Europ J Immunol 19:1393,1989.
13. Goller B, Kubbies M: UV-lasers for flow cytometric analysis: HeCd vs argon laser excitation. J Histochem Cytochem, 40:451,1992.
14. Rabinovitch PS: Analysis of percent responding cells in kinetic experiments by application of curve substraction. Cytometry Suppl. 5:138,1991.

11 Biochemical Parameters of Cell Function

G. ROTHE and G. VALET

11.1 Introduction

Cellular activation is characterized by early changes in functional parameters such as depolarization of the plasma membrane potential [1], increase in the cytosolic free calcium concentration [2], alkalinization of intracellular pH [3], increased mitochondrial charge [4], production and release of reactive oxidants [5], changes in cellular glutathione content [6], and activation and release of lysosomal proteases [7,8]. These parameters are sensitive indicators of stimulus-coupled biochemical processes even before protein synthesis, gene transcription, or cellular proliferation is induced.

Functional parameters are analyzed by flow cytometry following staining of vital cells with specific fluorescent or fluorogenic indicators. Intracellular ion concentrations such as the intracellular pH value and the cytosolic free calcium concentration are measured by changes in the fluorescence emission spectrum or intensity of ion-sensitive probes such as 2,3-dicyanohydroquinone (DCH), carboxy-seminaphthorhodafluor-1 (SNARF-1), indo-1, or fluo-3 (Fig. 1, Table 1). These dyes accumulate intracellularly by enzymatic cleavage of membrane permeable ester derivatives. The membrane potential of plasma or mitochondrial membrane can be measured by the accumulation of lipophilic fluorescent indicator dyes with a delocalized charge (Table 1). Cellular oxidants such as superoxide anion or hydrogen peroxide or antioxidants such as glutathione can be measured by the formation of specific fluorescent products with fluorogenic substrates (Figure 2, Table 1). Specific enzymatic activities such as lysosomal protease activities can be analyzed by the fluorescence generated by the intracellular cleavage of specifically N,N'-bis-peptide substituted rhodamine 110 (R110) derivatives (Figure 3, Table 1).

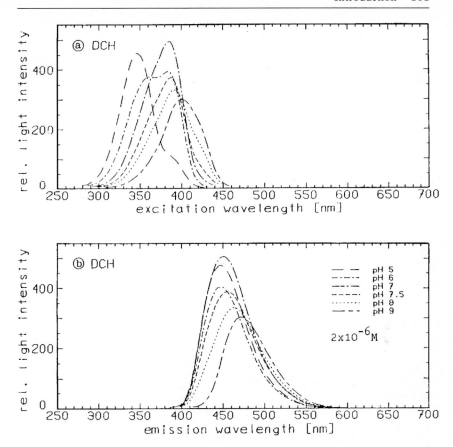

Fig. 1: Fluorescence excitation (a) and emission (b) spectra of 2,3-dicyanohydro-quinone in 0.15 M NaCl, 50 mM PO_4 buffers of pH 5 - 9

Table 1.

Indicator	Excitation	Principle of reaction	Ref.
Cytosolic free calcium			
Indo-1/AM	365 nm	Change of indo-1 green to blue fluorescence with increasing calcium concentration	[9-11]
Fluo-3/AM	488 nm	Increase of green fluorescence intensity with increasing calcium concentration	[12, 13]
Intracellular pH			
1,4-Diacetoxy-2,3-dicyanobenzene (ADB)	365 nm	Change of 2,3-dicyanohydroquinone (DCH) blue to green fluorescence with increasing pH value	[14-16]
SNARF-1/AM	488 nm or 514 nm	Change of SNARF-1 orange to red fluorescence with increasing pH value	[17]

Table 1. (continued)

Indicator	Excitation	Principle of reaction	Ref.
Membrane potential			
3,3'-Dihexyloxacarbo-cyanine iodide (DiOC6(3))	488 nm	Intracellular accumulation of green fluorescent cationic dye dependent on plasma membrane potential	[1, 18]
Bis-(1,3-dibutyl-barbiturate) trimethine oxonol	488 nm	Extracellular plasma membrane-associated accumulation of green fluorescent anionic dye dependent on membrane potential	[19, 20]
Rhodamine 123	488 nm	Mitochondrial accumulation of green fluorescent cationic dye dependent on mitochondrial membrane potential	[4]
Reactive oxidants			
Dihydrorhodamine 123	488 nm	Intracellular oxidation to green fluorescent rhodamine 123 by hydrogen peroxide in the presence of peroxidases	[21-23]
Hydroethidine	365 nm or 488 nm	Intracellular oxidation to red fluorescent ethidium bromide by superoxide anion or hydrogen peroxide and peroxidases	[24-26]
Reduced glutathione and SH-groups			
o-Phthaldialdehyde (OPT)	365 nm	Formation of a blue fluorescent product through coupling to both the primary amine and thiol moieties of glutathione	[27]
Monochlorobimane	365 nm	Glutathione-S-transferase-dependent formation of a blue fluorescent glutathione-adduct	[28-32]
Mercury orange	488 nm	Formation of an insoluble red fluorescent product with non-protein thiols	[33]
5-Chloromethyl-fluorescein diacetate	488 nm	Intracellular accumulation of green fluorescent hydrolytic product through coupling to sulfhydryl groups	[34]
Lysosomal proteinases			
N,N'-peptide substituted Rhodamine 110 derivatives	488 nm	Intracellular cleavage to green fluorescent monosubstituted and free rhodamine 110 by lysosomal serine or cysteine proteinases dependent on enzyme-specific peptide substituent	[35-37]

Lines of the high pressure mercury arc lamp or argon lasers used for fluorescence excitation

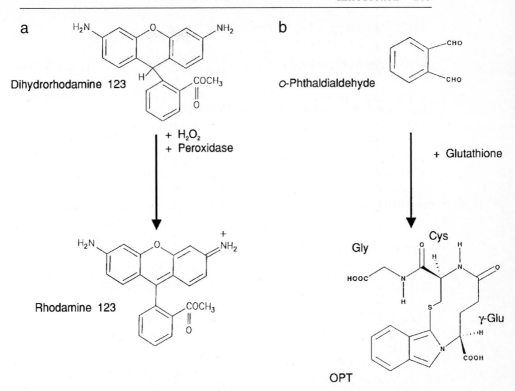

Fig. 2. Generation of fluorescent products by peroxidase-catalyzed oxidation of dihydrorhodamine 123 to rhodamine 123 **(a)** or coupling of o-phthaldialdehyde to the primary amine and thiol moieties of glutathione **(b)**

Rhodamine 110 diamide
(relative fluorescence < 10⁻⁴)

Rhodamine 110 monoamide
(relative fluorescence 0.1)

Rhodamine 110
(relative fluorescence 1)

Fig. 3. Generation of fluorescent monosubstituted rhodamine 110 and free rhodamine 110 by sequential proteolytic cleavage of N,N'-substituted rhodamine 110

11.2 Preparation of Fresh Cell Suspensions

11.2.1 Background

All functional assays require the preparation of fresh cell suspensions in a physiological environment. Saline solutions buffered by inert piperazine derivatives such as hydroxyethylpiperazine ethanesulfonic acid (HEPES) lead to better preservation of cell functions than bicarbonate-buffered solutions, as pH changes due to loss of CO_2 during incubation in ambient air are avoided. Divalent cations, glucose, amino acids, or proteins should be added for prolonged incubations, assays including cellular stimulation, or if cell losses are encountered, depending on the cell type (Sect. 11.2.3).

11.2.2 Material

- Hank's balanced salt solution (HBSS) without phenol red (Sigma, H 1387)
- Propidium iodide (PI; Serva 33671)

11.2.3 Method

1. Prepare a suspension of 10^7-10^8 cells/ml in either a buffered salt solution or a physiological medium, for example, autologous plasma. Avoid detrimental procedures such as lysis of erythrocytes or gradient centrifugations. Store cells on ice.
2. For specific staining dilute the cell suspension to 10^5 - 10^6 cells/ml with bicarbonate-free buffer such as HBSS supplemented with 10 mM HEPES instead of 10 mM bicarbonate (pH 7.35). A divalent cation-free HEPES-buffered saline (0.15 M NaCl, 5 mM HEPES, pH 7.35) may be used if neither cellular stimulation nor prolonged incubation is required. Use polypropylene tubes if cells are incubated at 37°C as polystyrene may lead to artifacts.
3. Counterstain dead cells before flow cytometry by adding 30 μM PI (stock 3 mM in HEPES-buffered saline) for 3 min on ice.
4. Analyze cells on flow cytometer recording data in list mode. Evaluate by gating on the viable cell cluster with linearization of logarithimically acquired data and quantitative calculation of means.

11.3 Cytosolic Free Calcium

11.3.1 Background

Cytosolic free Ca^{2+} is a universal second messenger, which is increased within seconds following cellular activation. In resting cells, the cytosolic free Ca^{2+} concentration is actively maintained low at 100 - 150 nM by a Ca^{2+} ATPase, compared to the 10^4-fold higher 1.3 mM free Ca^{2+} concentration in the extracellular space [2,38]. Upon surface receptor stimulation an increase in the cytosolic free Ca^{2+} concentration in nonmuscle cells typically occurs in three phases. First, receptor-coupled phospholipase C activity leads to an initial liberation of calcium from calciosomes, the intracellular calcium stores of nonmuscle cells. In the second phase an influx of calcium from the extracellular space is induced. This is followed within a few minutes by reuptake of Ca^{2+} by the calciosomes and export through the plasma membrane bound Ca^{2+} ATPase leading to the termination of the generalized increase in cytosolic Ca^{2+}. A sustained activation of calcium-dependent cellular responses is then maintained by locally increased calcium fluxes in the submembrane space only.

Under pathological conditions high cytosolic Ca^{2+} concentrations may be observed for a prolonged time, leading to excessive Ca^{2+} uptake by mitochondria. Cell death occurs finally through Ca^{2+}-induced activation of lipases, proteases, and endonucleases [39].

The cytosolic free calcium concentration can be measured intracellularly by the spectral shift of the dye indo-1 from green fluorescence in the absence of Ca^{2+} to blue fluorescence in the presence of Ca^{2+}. This allows precise determination of the calcium concentration independently of the cell size and cellular dye loading. First, indo-1 is accumulated intracellularly through intracellular cleavage of the membrane-permeable derivative indo-1/acetoxymethyl ester (AM) by esterases. The cellular blue to green fluorescence ratio is then analyzed as a measure of the intracellular free calcium concentration. A molar calibration curve can be obtained from the analysis of cells incubated in buffers of defined Ca^{2+} concentration in the presence of the Ca^{2+} ionophore ionomycin. Intracellular buffering of Ca^{2+} may lead to difficulties in calibrating concentrations of Ca^{2+} lower than the 100 - 150 nM typically found in resting cells. The spectral response of the flow cytometer when flushed with indo-1 and buffers of known Ca^{2+} concentrations should therefore also be recorded to compare the spectral Ca^{2+} calibration curve of the flow cytometer with the Ca^{2+} calibration curve of the cells.

In addition to differences in resting Ca^{2+} concentrations between different cell types, the kinetic response of intracellular Ca^{2+} to stimulation may be analyzed. This is achieved in three steps. First, cells loaded with indo-1 are analyzed at 37°C without stimulation. Then a stimulus is added, and the increase and decline in intracellular Ca^{2+} concentration are measured kinetically until a stable level is reached again. This typically occurs in 2 - 5 min. Finally, ionomycin is added to record the maximum indo-1 ratio in the presence of Ca^{2+} as a positive control.

11.3.2 Material

- Indo-1/AM (Calbiochem 402096) or fluo-3/AM (Molecular Probes C-1271)
- N-formyl-L-methionyl-L-leucyl-L-phenylalanine (fMLP; Sigma F 3506)
- Ionomycin (Calbiochem 407952)

11.3.3 Methods

Measurement of cytosolic free calcium with indo-1

1. Incubate 5 x 10^6 cells/ml in HEPES-buffered medium with 0.5 - 5 μM indo-1/AM at 37°C. An equilibrium of dye loading is reached within 20 - 30 min of incubation. Depending on the cell type, select the lowest concentration of substrate which results in a homogeneous fluorescence ratio prior to stimulation.
2. Add PI at 30 μM final concentration and analyze the blue fluorescence of indo-1/Ca^{2+} complexes using a 390- to 440-nm bandpass filter, and the blue-green fluorescence of Ca^{2+}-free indo-1 and the red PI fluorescence of dead cells using a 490-nm longpass filter with excitation at 351/363 nm by an argon laser or a high-pressure mercury arc lamp with a 360- to 370-nm bandpass filter.
3. For the analysis of stimulation-dependent changes in intracellular Ca^{2+}, add an appropriate stimulus, for example, fMLP ($10^{-8} M$) in the case of neutrophils, following an initial measurement for 30 s. Continue kinetic data acquisition for approximately 3 min.
4. Add ionomycin (2 μM) as a positive control for the maximal Ca^{2+} response of the cells.

If fluorescence excitation at 365 nm is not possible, for example, using argon laser based instruments which allow excitation at 488 nm only, the Ca^{2+}-sensitive fluorescein derivative fluo-3 may be used alternatively. Binding of Ca^{2+} does not change the fluorescence emission spectrum of fluo-3 but increases the green fluorescence intensity by 40-fold [12]. This allows sensitive detection of the kinetically fast relative increases in cytosolic free Ca^{2+} following stimulation (Fig. 4). Absolute Ca^{2+} concentrations, however, are difficult to measure with fluo-3 as the cellular fluorescence intensity is substantially affected by differences in enzymatic dye loading and cell size.

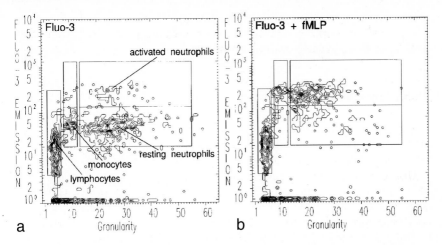

Fig. 4. Analysis of the kinetic increase in cytosolic free calcium in neutrophils and monocytes following stimulation with fMLP (10^{-8} *M*). Neutrophils and monocytes respond with a homogeneous increase in cellular fluo-3 green fluorescence 30 s after addition of fMLP **(b)** compared to the lower fluorescence before addition of the stimulus **(a)**.

Measurement of cytosolic free calcium with fluo-3

1. Incubate 5 x 10^6 cells/ml in HEPES-buffered medium with 0.5 - 2 μM fluo-3/AM at 37 °C. An equilibrium of dye loading is reached within 20 - 30 min of incubation. Depending on the cell type, select the lowest concentration of substrate which results in a homogeneous fluorescence prior to stimulation.
2. Add PI at 30 μM final concentration and analyze the green fluorescence of fluo-3 using a 515- to 535-nm bandpass filter, and the red PI fluorescence of dead cells using a 620-nm longpass filter with excitation by the 488-nm line of an argon laser or a high-pressure mercury arc lamp with a bandpass filter (range 470- 500 nm).
3. For the analysis of stimulation-dependent changes in intracellular Ca^{2+}, add an appropriate stimulus, for example, fMLP (10^{-8} *M*) in the case of neutrophils, following an initial measurement for 30 s. Continue kinetic data acquisition for approximately 3 min.
4. Add ionomycin (2 μM) as a positive control for the maximal Ca^{2+} response of the cells.

11.3.4 Alternative Methods

Recently, the new Ca^{2+}-sensitive dye fura red (Molecular Probes F-3020) has been introduced, the 488-nm excitable red fluorescence of which is decreased upon binding of Ca^{2+}. This dye should be useful for the simultaneous analysis of Ca^{2+} fluxes and the binding of fluorescein isothiocyanate

(FITC)-conjugated antibodies or the turnover of fluorogenic indicators which result in cellular green fluorescence, such as dihydrorhodamine 123 (DHR) or R110 based protease substrates.

11.3.5 Tips, Tricks, and Troubleshooting

If only low cellular accumulation of fluorescent indo-1 is reached, dye loading may be improved by addition of the non-ionic dispersing agent pluronic F-127 (Molecular Probes P-1572) [40].

Low changes in the cellular indo-1 fluorescence ratio even after addition of ionomycin may be due to incomplete hydrolysis of indo-1/AM. This may occur in some cell types resulting in Ca^{2+} insensitive fluorescence. A more complete hydrolysis may be reached by washing the cells and incubating for a prolonged time in a substrate-free medium prior to analysis.

A lack of a cellular response to stimulation despite a normal response following addition of ionomycin may be due to a low Ca^{2+} concentration extracellularly, for example, when media such as RPMI-1640 are used, or due to intracellular buffering of Ca^{2+} by the fluorescent indicator. This can occur due to high intracellular concentrations of the dyes in the micromolar range compared the 100-fold lower Ca^{2+} concentration and the low Kd values of the dyes for Ca^{2+}, for example, 250 nM in the case of indo-1. Therefore, the lowest substrate concentration should be used which results in a homogeneous fluorescence ratio of resting cells. No interaction of the dye loading procedure with the cellular sensitivity for stimulation may be shown by independent cellular responses such as chemotaxis, depolarization of the membrane potential, or an oxidative burst response in phagocytic cells.

11.4 Intracellular pH

11.4.1 Background

Intracellular pH is closely regulated in the range of 7.0 - 7.4 in eukaryotic cells [3]. The most common pH-regulatory mechanism is an amiloride-sensitive, electroneutral Na^+/H^+ antiport which uses the inward -directed Na^+ gradient for H^+ extrusion. Alkalinization of intracellular pH is induced by upregulation of the antiport following stimulation by hormones or growth factors. In addition, pH-regulatory anion antiport mechanisms such as an Cl^-/HCO_3^- antiport seem to be present in most cells. The regulation of the intracellular pH is of major importance for cellular activation as increases in intracellular pH are associated with an increased metabolic activity of cells. This is mediated through increases in phosphofructokinase activity, which is rate limiting for glycolysis, protein synthesis, and DNA polymerase activity. Thus a rapid increase in intracellular pH may be important for the transition of cells into the S phase. The cytoplasmic pH value can be measured intracellularly through the spectral shift of dyes such as 365-nm excitable DCH or the 488-nm excitable SNARF-1 to a longer wavelength fluorescence emission with increasing pH value. In the first step, DCH or SNARF-1 is accumulated intracellularly by intracellular cleavage of the membrane-permeable derivatives 1,4-diacetoxy-2,3-dicyanobenzene (ADB) or SNARF-1/AM by esterases. The cellular blue to green (DCH) or red to orange (SNARF-1) fluorescence ratio is then analyzed as a measure of the intracellular pH value. A pH calibration curve can be obtained from the analysis of cells incubated in buffers of defined pH value with equilibration of the intracellular pH by addition of the ionophore nigericin in the presence of a high K^+ concentration [41]. In addition, the spectral response of the flow cytometer may be recorded when flushed with DCH or SNARF-1 containing buffers of known pH values [14].

In addition to differences in resting intracellullar pH value between different cell types, the response of the intracellular pH to stimulation may be analyzed (Fig. 5). The stimulation-dependent activation of the Na^+/H^+ antiport typically leads to prolonged cytoplasmic alkalinization. This response can be analyzed by addition of the stimulus during the SNARF-1 loading procedure at 37°C (Sect. 11.4.3). DCH, in contrast to SNARF-1, is efficiently accumulated intracellularly at a low temperature. Therefore, using DCH, cells may be alternatively preincubated at 37°C with the stimulus in the absence of the pH indicator followed by stopping of the cellular response by transfer onto ice and loading with DCH on ice or at room temperature (Sect. 11.4.3).

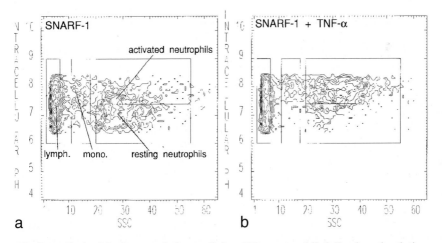

Fig. 5. Analysis of the increase in intracellular pH in neutrophils following stimulation with TNF-α (1 ng/ml). Neutrophils respond with a homogeneously alkaline intracellular pH after addition of TNF-α **(b)** compared to a heterogeneous intracellular pH before addition of the stimulus **(a)**

11.4.2 Material

<table>
<tr><td>Measurement
with
SNARF-1</td><td>

– SNARF-1/AM (Molecular Probes C-1271).
– fMLP (Sigma F 3506)
– Tumor necrosis factor-α (TNF-α; Sigma T-0517)
– Nigericin (Molecular Probes N-1495)
– High K^+ buffers: 140 mM KCl, 10 mM 2-[N-morpholino]ethanesulfonic acid (MES, Sigma M-8250), 10 mM HEPES (Serva 25245), adjusted with NaOH to pH values 6.40, 6.80, 7.20, 7.60, 8.00, and 8.40
</td></tr>
</table>

<table>
<tr><td>Measurement
with DCH</td><td>

– ADB (Calbiochem 266707)
– fMLP (Sigma F 3506)
– TNF-α (Sigma T-0517)
– Nigericin (Molecular Probes N-1495)
– High K^+ buffers: 140 mM KCl, 10 mM 2-[N-morpholino]ethanesulfonic acid (MES, Sigma M-8250), 10 mM HEPES (Serva 25245), adjusted with NaOH to pH values 6.40, 6.80, 7.20, 7.60, 8.00, and 8.40
</td></tr>
</table>

11.4.3 Methods

Measurement of intracellular pH with SNARF-1

1. Incubate 5 x 10^6 cells/ml in HEPES-buffered medium with 0.2 - 1 μM SNARF-1/AM at 37 °C. An equilibrium of dye loading is reached within 20 - 30 min of incubation. Depending on the cell type, select the lowest concentration of substrate which results in a homogeneous fluorescence ratio in resting cells.
2. For the analysis of stimulation-dependent changes in intracellular pH, incubate for the final 15 min in the presence of appropriate stimuli, for example, fMLP (10^{-8} M) or TNF-α (1 ng/ml) in the case of neutrophils.
3. Add PI at 30 μM final concentration and analyze the orange fluorescence of acidic SNARF-1 using a 575- to 595-nm bandpass filter, and the red fluorescence of basic SNARF-1 and the PI fluorescence of dead cells using a 620-nm longpass filter with excitation by the 488-nm line of an argon laser or a high-pressure mercury arc lamp with a bandpass filter (range 470 - 500 nm).
4. For the generation of a calibration curve incubate cells with SNARF-1/ AM as above. Divide the sample into six aliquots and spin down at 60 g and 4 °C for 5 min. Resuspend for 5 min at room temperature in the high-K^+ calibration buffers (pH 6.40, 6.80, 7.20, 7.60, 8.00, 8.40) supplemented with 10 μM nigericin followed by flow cytometric analysis.

Measurement of intracellular pH with DCH

1. Incubate 5 x 10^6 cells/ml in HEPES-buffered medium without or with appropriate stimuli, for example, fMLP (10^{-8} M) or TNF-α (1 ng/ml) in the case of neutrophils. Stop incubation by transfer of cells onto ice.
2. Incubate the cells at room temperature or on ice with 10 - 200 μM ADB. An equilibrium of dye loading is reached within 5 - 15 min of incubation.

Depending on the cell type select the lowest concentration of substrate which results in a homogeneous fluorescence ratio in resting cells.

3. Add PI at 30 μM final concentration and analyze the blue fluorescence of acidic DCH using a 390- to 440-nm bandpass filter, and the green to red fluorescence of basic DCH and the PI fluorescence of dead cells using a 490-nm longpass filter with excitation at 351/363 nm by an argon laser or a high-pressure mercury arc lamp with a 360- 370-nm bandpass filter.

4. For the generation of a calibration curve incubate cells with ADB as above. Divide the sample into six aliquots and spin down at 60 g and 4 °C for 5 min. Resuspend for 5 min at room temperature in the high-K^+ calibration buffers (pH 6.40, 6.80, 7.20, 7.60, 8.00, 8.40) supplemented with 10 μM nigericin

11.4.4 Tips, Tricks, and Troubleshooting

Cytosol has a high buffering capacity for intracellular pH. Therefore, the additional buffering of the intracellular pH value by the intracellular indicator usually does not significantly affect the pH measurement.

A lack of a cellular response to stimulation despite pH-sensitive cellular fluorescence as observed by calibration with nigericin may be due to the simultaneous activation of H^+-extruding antiports and metabolic activity which leads to the intracellular generation of H^+. Na^+/H^+ antiport activity can be shown in this case by pH changes following preincubation with the specific Na^+/H^+ antiport inhibitor amiloride (Sigma A 7410).

11.5 Oxidative Burst Activity of Phagocytic Cells

11.5.1 Background

Phagocytic cells are equipped with an enzymatic cascade for the production and extracellular release of the reactive oxidants superoxide anion, hydrogen peroxide, and hypochlorous acid during the respiratory burst [5]. The secretion of these oxidants is important for the killing of microorganisms but is also associated with inflammatory tissue destruction in diseases such as sepsis, posttraumatic organ failure, or autoimmune diseases [8]. The first step in the oxidative burst cascade is the one-electron reduction of molecular oxygen to superoxide anion by a membrane-bound nicotinamide adenine dinucleotide phosphate, reduced, (NADPH) oxidase. Dismutation of superoxide anion then results in the formation of hydrogen peroxide, which is further converted by lysosomal myeloperoxidase to hypochlorous acid and other stable oxidants.

Formation of the reactive oxidants superoxide anion and hydrogen peroxide during the oxidative burst is measured intracellularly by the oxidation of membrane-permeable fluorogenic substrates such as DHR or hydroethidine (HE) to fluorescent products. The intracellular oxidation of nonfluorescent DHR to green fluorescent rhodamine 123 (Fig. 2) by hydrogen peroxide and peroxidases is the most sensitive method for the analysis of the oxidative burst response [21,22,42,43]. This high sensitivity is reached by the intracellular retention of the positively charged fluorescent product rhodamine 123 at mitochodrial binding sites. The blue fluorescent HE, in contrast to DHR, can be oxidized to red fluorescent ethidium bromide (EB) already by superoxide anion, the primary product of the oxidative burst response [24]. The phagocytic cells, i.e., neutrophils, monocytes, or macrophages, are first preincubated with the fluorogenic substrates at 37°C. An oxidative burst response is then induced by addition of a physiological stimulus, such as the bacterial peptide fMLP with or without response modifiers such as the cytokine TNF-α (Fig. 6). Stimuli which are independent of the cellular signal transduction, such as the tumor promoter phorbol 12-myristate 13-acetate (PMA), may be used as a positive control or for the analysis of the cellular expression of the oxidative burst enzymes. The incubation is stopped by transferring cells onto ice. The cellular accumulation of the fluorescent oxidation products is then analyzed on the flow cytometer. As the cellular fluorescence is stable for about 90 min on ice, cells may be subsequently labelled with mAbs.

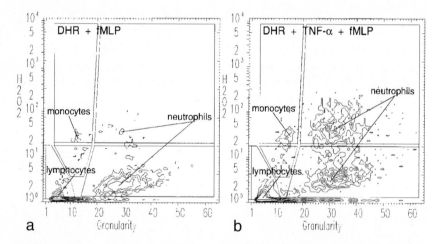

Fig. 6. Analysis of the oxidative burst response of neutrophils following combined stimulation with TNF-α (10 ng/ml) and fMLP (10^{-7} M). Neutrophils show only a low oxidative burst response to stimulation by fMLP alone **(a)**. Preincubation with TNF-α, which itself also only induces a low oxidative burst response leads to a high and heterogeneous oxidative burst response of a subpopulation of neutrophils to subsequent fMLP stimulation **(b)**

11.5.2 Material

- DHR (Molecular Probes D-632).
- HE (Molecular Probes D-1168; distributed as dihydroethidium)
- fMLP (Sigma F 3506)
- TNF-α (Sigma T-0517)
- PMA (Sigma P 8139)

11.5.3 Method

1. Incubate 5×10^6 cells/ml in HEPES-buffered medium with 1 µM DHR (stock 1 mM in N,N-dimethylformamide, DMF) or 10 µM HE (stock 10 mM in DMF) at 37°C for 5 min.
2. Incubate for 15 - 30 min with appropriate stimuli, for example, fMLP (10^{-6} M) or PMA (10^{-7} M) in the case of neutrophils. The cellular „priming" by incomplete stimuli such as TNF-α (10 ng/ml) may be analyzed by 5-min preincubation with the cytokine followed by incubation for 15 min a low concentration of fMLP (10^{-7} M).
3. Add PI at 30 µM final concentration and analyze the green fluorescence of rhodamine 123 using a 515- to -535-nm bandpass filter, and the PI red fluorescence of dead cells or the EB red fluorescence of HE-stained cells using a 620-nm longpass filter using excitation with the 488-nm line of an argon laser or a high-pressure mercury arc lamp with a bandpass filter (range 470- 500 nm).
4. For simultaneous immunofluorescence spin down the sample and incubate with antibodies conjugated with R-phycoerythrin or preferably R-phycoerythrin-Cy5 tandem detection systems in the case of DHR. Use FITC-labelled antibodies in the case of HE.

11.5.4 Alternative Methods

The intracellular oxidation of 2´,7´-dichlorofluorescin (DCFH) to green fluorescent 2´,7´-dichlorofluorescein was used in an earlier flow cytometric assay for the oxidative burst response of phagocytic cells [45]. DCFH oxidation in comparison to the intracellular oxidation of DHR results in a far lower cellular fluorescence of stimulated phagocytic cells, making impossible the analysis of cellular responses to low amounts of physiological stimuli [43]. The assay is further complicated by a high background fluorescence also in cells which are not capable of oxidative burst activity, such as lymphocytes and the need to accumulate the fluorogenic substrate in a first step through a hydrolytic intracellular cleavage of the membrane-permeable derivative 2´,7´-dichlorofluorescin diacetate.

11.5.5 Tips, Tricks, and Troubleshooting

Only low cellular responses to stimulation may be due to the presence of high [> 0.1% (v/v)] concentrations of organic solvents such as dimethyl-sulfoxide (DMSO) or DMF used in the reagent stock solutions, which both act as scavenger for reactive oxidants and interfere with cellular stimulation.

High fluorescence in nonphagocytic cells such as lymphocytes indicates spontaneous oxidation of the fluorogenic substrates, which are unstable when exposed to excessive light or stored at higher temperature.

Specificity of the intracellular accumulation of the fluorescent products for the oxidative burst response may be shown by preincubating the cells with 100 μM diphenyl iodonium (Aldrich-Chemie D20,908-2, stock 100 mM in DMSO) as a specific inhibitor of the NADPH oxidase of phagocytic cells [44]. Hydrogen peroxide (1 mM; Merck 7210) may be used as positive control to show sensitivity of the substrates for intracellular oxidation.

11.6 Reduced Glutathione

11.6.1 Background

The tripeptide glutathione (γ-Glu-CysH-Gly; GSH) is both the most prevalent cellular thiol and the most abundant low molecular weight peptide, being typically present intracellularly at levels of 0.1 - 10 mM [46]. GSH acts as the major cellular oxidant scavenger in either a GSH peroxidase or a GSH S-transferase catalyzed reaction. The analysis of cellular GSH levels is therefore interesting as an indicator of cellular sensitivity and response to oxidative stress during inflammatory reactions, cytostatic drug therapy with alkylating reagents, and radiotherapy [6,47].

Intracellular GSH levels can be measured intracellularly by the spontaneous formation of specifically fluorescent adducts with the nonfluorescent membrane-permeable substrate o-phthaldialdehyde (OP) [27]. When coupling to GSH, OPT reacts with both the primary amine and thiol moieties of GSH forming a tricyclic isoindole structure (Fig. 2) [48,49] which has a longer wavelength fluorescence than bicyclic OPT protein adducts. Specificity of the cellular 475- to 495- nm fluorescence for GSH-OPT adducts can be further controlled by a nearly complete inhibition following the blocking of cellular sulfhydryl groups through preincubation with HgCl$_2$ or inhibition of glutathione synthesis by buthionine sulfoximine.

11.6.2 Material

– OPT (Sigma P-0657).
– HgCl$_2$ (Merck 4404)

11.6.3 Method

1. Incubate 5×10^6 cells/ml in HEPES-buffered medium with $100 - 500\,\mu M$ OPT (stock 1000-fold in DMF). Cellular GSH specific fluorescence is reached in equilibrium within 5 min independently of the temperature (ice, room temperature, or 37 °C).
2. Add PI at $30\,\mu M$ final concentration and analyze the blue fluorescence of bicyclic GSH protein adducts at 395-415 nm, the green fluorescence of the tricyclic OPT-GSH adduct using a 475- to 495-nm bandpass filter, and the red PI fluorescence of dead cells using a 620-nm longpass filter using excitation with the 365-nm line of an argon laser or a high-pressure mercury arc lamp with 360- to 370-nm bandpass filter.
3. For control of the specificity of the reaction preincubate with $1\,mM$ HgCl$_2$ for 5 min. This should result in a more than 95% reduction in the fluorescence of OPT-GSH adducts.

11.6.4 Alternative Methods

An alternative method is based on the enzymatically faster reaction of the sulfhydryl derivatizing reagent monochlorobimane with GSH in the presence of GSH S-transferase than with other sulfhydryl groups [28,29]. As the specificity of the method is derived only from the kinetical difference of the two reactions, the method is highly temperature dependent, and kinetic stopping of the incubation by washing of the cells is required. Monochlorobimane may, however, be an interesting substrate in the study of the cellular expression of different GSH S-transferase isozymes, which has been shown substantially to affect the intracellular staining reaction [30-32].

Methods detecting intracellular GSH with 488-nm excitation either lack specificity for GSH compared to other thiols in the case of the substrate 5-chloromethylfluorescein diacetate [34] or require staining in acetone in the case of the substrate mercury orange [33], restricting the applicability to analytical purposes.

11.6.5 Tips, Tricks, and Troubleshooting

A 488-nm excitable cellular green fluorescence may develop in some cell types when using high OPT concentrations and interfere with the detection of FITC-conjugated antibodies in dual-laser analysis. This may be avoided by using a lower OPT concentration still above the minimal saturating OPT concentration as obtained from a titration experiment.

11.7 Lysosomal Cysteine and Serine Proteinases

11.7.1 Background

Cellular endopeptidases are comprised of four classes of enzymes defined by the action of class-specific inhibitors, serine proteinases inhibitable by diisopropylphosphofluoridate (DFP), cysteine proteinases inhibitable by E-64, aspartic proteinases inhibitable by pepstatin, and metalloproteinases inhibitable by phenanthroline [50]. Due to a wide substrate specificity these enzymes are involved in intracellular protein turnover as well as extracellular protein degradation during inflammation or metastasis [51]. Cell lineage dependent differences in the cellular expression of lysosomal proteases, for example, expression of the serine proteinases elastase and cathepsin G in neutrophils in contrast to expression of the cysteine proteinases cathepsin B and L in monocytes and macrophages furthermore make them interesting as markers of cellular differentiation (Fig. 7).

Lysosomal proteinase activity can be analyzed intracellularly by the conversion of nonfluorescent and membrane-permeable , specifically bis-substituted R110 peptide derivatives to green fluorescent monosubstituted R110 and free R110 in a two-step reaction (Fig. 2) [35-37]. Cells are incubated kinetically at 37°C with appropriate substrates such as (Z-Arg-

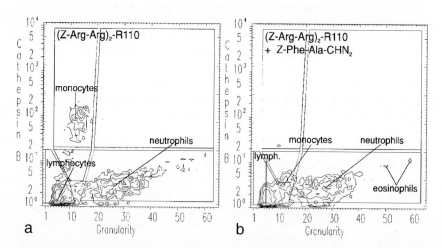

Fig. 7. Analysis of the lysosomal cysteine proteinases cathepsin B and L of monocytes. Monocytes develop a high green fluorescence during incubation with (Z-Arg-Arg)2-R110 in contrast to no significant accumulation of fluorescence by neutrophils or lymphocytes **(a)**. Preincubation of the cells with the cysteine proteinase inhibitor Z-Phe-Ala-CHN$_2$ leads to a nearly complete inhibition of the reaction **(b)** indicating specificity of the reaction

Arg)$_2$-R110 for the cysteine proteinases cathepsin B and L of monocytes or (Z-Ala-Ala)$_2$-R110 for the serine proteinase elastase of neutrophils. The specificity of the reaction is shown by the inhibition of the reaction through preincubation with specific inhibitors such as (Z-Phe-Ala-CHN$_2$ for cathepsin B and L [52] or DFP for neutrophil elastase [53]. As the cellular fluorescence is stable for about 90 min on ice, cells may be subsequently labelled with mAbs.

11.7.2 Material

– (Z-Arg-Arg)$_2$-R110 (synthesized in analogy to Leytus et al. [36,37]
– Z-Phe-Ala-CHN$_2$ (Bachem N-1040)
– (Z-Ala-Ala)$_2$-R110 (synthesized in analogy to Leytus et al. [36,37]
– DFP (Aldrich D12,600-4)

11.7.3 Method

1. Preincubate 5 x 10^6 cells/ml in HEPES-buffered medium with or without specific proteinase inhibitor such as 100 μM Z-Phe-Ala-CHN$_2$ (stock 100 mM in DMSO) or 1 mM DFP (stock 1 M in DMSO) at 37°C for 10 min.
2. Incubate for 20 min with appropriate proteinase substrate such as 4 μM (Z-Arg-Arg)$_2$-R110 for the cathepsin B/L of monocytes or 4 μM (Z-Ala-Ala)$_2$-R110 for the elastase of neutrophils
3. Add PI at 30 μM final concentration and analyze the green fluorescence of R110 using a 515- to 535-nm bandpass filter, and the red PI fluorescence of dead cells using a 620-nm longpass filter using excitation with 488-nm line of an argon laser or a high-pressure mercury arc lamp with a bandpass filter (range 470- 500 nm).
4. For simultaneous immunofluorescence spin down the sample and incubate with antibodies conjugated with R-phycorerythrin of preferably R-phycoerythrin-Cy5 tandem detection systems.

11.7.4 Alternative Methods

Earlier methods for the intracellular analysis of protease activities were based on the intracellular cleavage of N-acyl derivatives of 4-methoxy-2-napthylamine. The highly diffusible product 4-methoxy-2-naphthylamine was trapped inside the cell by coupling with 5-nitrososalicylaldehyde, yielding a yellow fluorescent, crystalline product [54-57]. These methods are restricted to the analysis of proteases with an acidic pH optimum, as incubation at acidic pH is necessary to promote intracellular coupling of the 5-nitrososalicylaldehyde to the product. Furthermore, a high background is induced by the staining of proteins with the coupling reagent.

11.7.5 Tips, Tricks, and Troubleshooting

Cell death selectively occurring in samples incubated with the proteinase substrates but not in samples preincubated with the inhibitors prior to incubation with the proteinase substrates may be a result of the local accumulation of high amounts of the product R110 inside cellular lysosomes. This can be avoided by reducing the substrate concentration or time of incubation.

References

1. Shapiro HM (1981) Flow cytometric probes of early events in cell activation. Cytometry 1:301-312
2. Rasmussen H, Rasmussen JE (1990) Calcium as intracellular messenger: from simplicity to complexity. Curr Top Cell Regul 31:1-109
3. Madshus IH (1988) Regulation of intracellular pH in eukaryotic cells. Biochem J 250:1-8
4. Chen LB (1988) Mitochondrial membrane potential in living cells. Ann Rev Cell Biol 4:155-181
5. Clark RA (1990) The human neutrophil respiratory burst oxidase. J Infect Dis 161:1140-1147
6. Bilzer M, Lauterburg BH (1991) Glutathione metabolism in activated human neutrophils: stimulation of glutathione synthesis and consumption of glutathione by reactive oxygen species. Eur J Clin Invest 21:316-322
7. Tschesche H, Macartney HW (1981) A new principle of regulation of enzymic activity. Activation and regulation of human polymorphonuclear leukocyte collagenase via disulfide-thiol exchange as catalysed by the glutathione cycle in a peroxidase-coupled reaction to glucose metabolism. Eur J Biochem 120:183-190
8. Weiss SJ (1989) Tissue destruction by neutrophils. N Engl J Med 320:365-376
9. Grynkiewicz G, Poenie M, Tsien RY (1985) A new generation of Ca^{2+} indicators with greatly improved fluorescence properties. J Biol Chem 260:3440-3450
10. Valet G, Raffael A, Rüssmann L (1985) Determination of intracellular calcium in vital cells by flow-cytometry. Naturwiss 72:600-602
11. Rabinovitch PS, June CH, Grossmann A, Ledbetter JA (1986) Heterogeneity among T-cells in intracellular free calcium responses after mitogen stimulation with PHA or anti-CD3. Simultaneous use of indo-1 and immunofluorescence with flow cytometry. J Immunol 137:952-961
12. Minta A, Kao JPY, Tsien RY (1989) Fluorescent indicators for cytosolic calcium based on rhodamine and fluorescein chromophores. J Biol Chem 264:8171-8178
13. Kao JPY, Harootunian AT, Tsien RY (1989) Photochemically generated cytosolic calcium pulses and their detection by fluo-3. J Biol Chem 264:8179-8184
14. Valet G, Raffael A, Moroder L, Wünsch E, Ruhenstroth-Bauer G (1981) Fast intracellular pH determination in single cells by flow cytometry. Naturwiss 68:265-266
15. Musgrove E, Rugg C, Hedley D (1986) Flow cytometric measurement of cytoplasmic pH: a critical evaluation of available fluorochromes. Cytometry 7: 347-357
16. Cook JA, Fox MH (1988) Intracellular pH measurements using flow cytometry with 1,4-diacetoxy-2,3-dicyanobenzene. Cytometry 9:441-447
17. Whitaker JE, Haugland RP, Prendergast FG (1991) Spectral and photophysical studies of benzo[c]xanthene dyes: Dual emission pH sensors. Anal Biochem 194:330-344
18. Sims PJ, Waggoner AS, Wang CH, Hoffman JF (1974) Studies on the mechanism by which cyanine dyes measure membrane potential in red blood cells and phosphatidylcholine vesicles. Biochemistry 13:3315-3330

19. Rink TJ, Montecuco C, Hesketh TR, Tsien RY (1980) Lymphocyte membrane potential assessed with fluorescent probes. Biochim Biophys Acta 595:15-30

20. Wilson AH, Chused TM (1985) Lymphocyte membrane potential and Ca^{2+}-sensitive potassium channels described by oxonol dye fluorescence measurements. J Cell Physiol 125:72-81

21. Rothe G, Oser A, Valet G (1988) Dihydrorhodamine 123: a new flow cytometric indicator for respiratory burst activity in neutrophil granulocytes. Naturwiss 75:354-355

22. Lund-Johansen F, Olweus J, Aarli A, Bjerknes R (1990) Signal transduction in human monocytes and granulocytes through the PI-linked antigen CD14. FEBS Letters 273:55-58

23. Roesler J, Hecht M, Freihorst J, Lohmann-Matthes ML, Emmendörffer A (1991) Diagnosis of chronic granulomatous disease and of its mode of inheritance by dihydrorhodamine 123 and flow microcytofluorometry. Eur J Pediatr 150:161-165

24. Rothe G, Valet G (1990) Flow cytometric analysis of respiratory burst activity in phagocytes with hydroethidine and 2',7'-dichlorofluorescin. J Leukocyte Biol 47:440-448

25. Kobzik L, Godleski JJ, Brain JD (1990) Selective down-regulation of alveolar macrophage oxidative response to opsonin-independent phagocytosis. J Immunol 144:4312-4319

26. Perticarari S, Presani G, Mangiarotti MA, Banfi E (1991) Simultaneous flow cytometric method to measure phagocytosis and oxidative products by neutrophils. Cytometry 12:687-693

27. Treumer J, Valet G (1986) Flow-cytometric determination of glutathione alterations in vital cells by o-phthaldialdehyde (OPT) staining. Exp Cell Res 163:518-524

28. Rice GC, Bump EA, Shrieve DC, Lee W, Kovacs M (1986) Quantitative analysis of cellular glutathione by flow cytometry utilizing monochlorobimane: some applications to radiation and drug resistance in vitro and in vivo. Cancer Res 46:6105-6110

29. Shrieve DC, Bump EA, Rice GC (1988) Quantitative analysis of cellular glutathione among cells derived from a murine fibrosarcoma or a human renal carcinoma detected by low cytometric analysis. J Biol Chem 263:14107-14114

30. Ublacker GA, Johnson JA, Siegel FL, Mulcahy RT (1991) Influence of glutathione S-transferases on cellular glutathione determination by flow cytometry using monochlorobimane. Cancer Res 51:1783-1788

31. Cook JA, Iype SN, Mitchell JB (1991) Differential sensitivity of monochlorobimane for isozymes of human and rodent glutathione S-transferases. Cancer Res 51:1606-1612

32. Puchalski RB, Manoharan TH, Lathrop AL, Fahl WE (1991) Recombinant glutathione S-transferase (GST) expressing cells purified by flow cytometry on the basis of a GST-catalyzed intracellular conjugation of glutathione to monochlorobimane. Cytometry 12:651-655

33. O'Connor JE, Kimler BF, Morgan MC, Tempas KJ (1988) A flow cytometric assay for intracellular nonprotein thiols using mercury orange. Cytometry 9:529-532

34. Poot M, Kavanagh TJ, Kang HCh, Haugland RP, Rabinovitch PS (1991) Flow cytometric analysis of cell cycle-dependent changes in cell thiol level by combining a new laser dye with Hoechst 33342

35. Rothe G, Klingel S, Assfalg-Machleidt I, Machleidt W, Zirkelbach Ch, Mangel WF, Valet G (1992) Flow cytometric analysis of protease activities in vital cells. Biol Chem Hoppe-Seyler (Suppl.): in press

36. Leytus SP, Patterson WL, Mangel WF (1983) New class of sensitive and selective fluorogenic substrates for serine proteinases. Amino acid and dipeptide derivatives of rhodamine. Biochem J 215:253-260

37. Leytus SP, Melhado LL, Mangel WF (1983) Rhodamine-based compounds as fluorogenic substrates for serine proteinases. Biochem J 209:299-307

38. DiVirgilio F, Stendahl O, Pittet D, Lew PD, Pozzan T (1990) Cytoplasmic calcium in phagocyte activation. Curr Topics Membranes Transport 35:179-205

39. Orrenius S, McConkey DJ, Bellomo G, Nicotera P (1989) Role of Ca^{2+} in toxic cell killing. TIPS 10:281-285
40. Poenie M, Alderton J, Steinhardt R, Tsien R (1986) Calcium rises briefly and throughout the cell at the onset of anaphase. Science 233:886-889
41. Thomas JA, Buchsbaum RN, Zimniak A, Racker E (1979) Intracellular pH measurements in Ehrlich ascites tumor cells utilizing spectroscopic probes generated in situ. Biochemistry 18:2210-2218
42. Rothe G, Valet G (1990) Flow cytometric characterization of oxidative processes in neutrophils and monocytes with dihydrorhodamine 123, 2',7'-dichlorofluorescin and hydroethidine In: Burger G, Oberholzer M, Vooijs GP (eds) Advances in analytical cellular pathology. Elsevier, Amsterdam, pp 313-314
43. Rothe G, Emmendörffer A, Oser A, Roesler J, Valet G (1991) Flow cytometric measurement of the respiratory burst activity of phagocytes using dihydrorhodamine 123. J mmunol Methods 138:133-135
44. Cross AR (1987) The inhibitory effects of some iodonium compounds on the superoxide generating system of neutrophils and their failure to inhibit diaphorase activity. Biochem Pharmacol 36:489-493
45. Bass DA, Parce JW, Dechatelet LR, Szejda P, Seeds MC, Thomas M (1983) Flow cytometric studies of oxidative burst formation by neutrophils: a graded response to membrane stimulation. J Immunol 130:1910-1917
46. Meister A (1988) Glutathione metabolism and its selective modification. J Biol Chem 263:17205-17208
47. Morrow CS, Cowan KH (1990) Glutathione S-transferases and drug resistance. Cancer Cells 2:15-22
48. Neuschwander-Tetri BA, Roll FJ (1989) Glutathione measurement by high-performance liquid chromatography separation and fluorometric detection of the glutathione-orthophthalaldehyde adduct. Anal Biochem 179:236-241
49. Morineau G, Azoulay M, Frappier F (1989) Reaction of o-Phthalaldehyde with amino acids and glutathione. Application to high-performance liquid chromatography determination. J Chromatography 467:209-216
50. Kirschke H, Barrett AJ (1987) Chemistry of lysosomal proteases. In: Glaumann H, Ballard FJ (eds) Lysosomes: their role in protein breakdown. Academic Press, London, pp 193-238
51. Assfalg-Machleidt I, Jochum M, Nast-Kolb D, Siebeck M, Billing A, Joka T, Rothe G, Valet G, Zauner R, Scheuber HP, Machleidt W (1990) Cathepsin B - indicator for the release of lysosomal cysteine proteinases in severe trauma and inflammation. Biol Chem Hoppe-Seyler 371 (Suppl.):211-222
52. Green DJ, Shaw E (1981) Peptidyl diazomethyl ketones are specific inactivators of thiol proteinases. J Biol Chem 256:1923-1928
53. Powers JC (1986) Serine proteases of leukocyte and mast cell origin: substrate specificity and inhibition of elastase, chymases, and tryptases. Adv Inflammation Res 11:145-157
54. Dolbeare FA, Smith RE (1977) Flow cytometric measurement of peptidases with use of 5-nitrososalicylaldehyde and 4-methoxy-beta-naphthylamine derivatives. Clin Chem 23:1485-1491
55. Murphy RF (1985) Analysis and isolation of endocytic vesicles by flow cytometry and sorting: demonstration of three kinetically distinct compartments involved in fluid-phase endocytosis. Proc Natl Acad Sci USA 82:8523-8526
56. Krepela E, Bártek J, Skalková D, Vicar J, Rasnick D, Taylor-Papadimitriou J, Hallowes RC (1987) Cytochemical and biochemical evidence of cathepsin B in malignant, transformed and normal breast epithelial cells. J Cell Sci 87:145-154
57. Van Noorden CJF, Vogels IMC, Smith RE (1989) Localization and cytophotometric analysis of cathepsin B activity in unfixed and undecalcified cryostat sections of whole rat knee joints. J Histochem Cytochem 37:17-624

12 Molecular Analysis of Transcriptional Control: The FACS-Galactosidase Assay

S. JUNG

12.1 Background

The FACS-galactosidase (FACS-Gal) assay measures *Escherichia coli* *lacZ*-encoded β-galactosidase activity in eukaryotic cells. Based on analysis by flow cytometry, the enzyme assay allows the simultaneous use of even a single, stably integrated *lacZ* gene as a reportergene for functional analysis of transcriptional control elements. Apart from providing good statistics due to the high number of cells analyzed, *lacZ* expression can be measured quantitatively on a per cell basis. Also, *lacZ* expression can be correlated with other cellular parameters such as immunophenotype. Cells can be sorted alive according to *lacZ* expression, from less than 5 to more than 50 000 enzyme molecules per cell, for further cellular and molecular studies.

The principle of the FACS-Gal assay is that the cells to be analyzed are loaded with the fluorogenic substrate fluorescein-β-di-galactoside (FDG) by using a short hypotonic shock (figure). Hydrolytic cleavage of FDG by b-galactosidase frees fluorescein that is locked inside the cells if they are kept on ice, thus „freezing" the membrane. Addition of the competitive inhibitor of FDG, phenylethyl-β-*D*-thiogalactoside (PETG) slows the reaction down and stops it before the substrate FDG is exhausted.

12.2 Material

- FDG (Molecular Probes, Eugene, Oregon; 1179) dissolved in ethanol/dimethylsulfoxide (DMSO; 1:1), stored as 2 mM stock solution in 98% H_2O, 1% DMSO, 1% ethanol (can be freeze-thawed)
- Staining medium: RPMI deficient medium, 5% fetal calf serum (FCS), 10 mM hydroxyethylpiperazine ethanesulfonic acid (HEPES) (phosphate-buffered saline, 5% FCS, 10 mM HEPES has been shown to work as well)

– PETG (Sigma P4902) dissolved in H_2O to obtain 50 mM solution, sterilized by filtration and stored frozen in aliquots (can be freeze-thawed)

12.3 Method

Cell preparation
1. Grow cells to be analyzed under exponential conditions.
2. Harvest by trypsinization of adherent cells.
3. Spin cells and resuspend at 10^7/ml in staining medium.

Assay protocol
1. Place cells (in 50 µl medium) in 37°C water bath for 10 min.
2. Add 50 µl prewarmed (37^{0}C) 2 mM FDG solution in H_2O, mix rapidly and thoroughly.
3. Return to 37^{0}C water bath for exactly 1 min.
4. Stop FDG loading at the end of 1 min by adding 1 ml ice-cold staining medium.
5. Keep the cells on ice until flow analysis, gently resuspend them from time to time (use cooling devices of flow cytometer, if available).

12.4 Tips, Tricks, and Troubleshooting

Free fluorescein in the FDG solution
Fluorescein derived from spontaneous hydrolysis of FDG raises the background. This fluorescein can be „removed" by photobleaching. We hold the 2 mM solution for about 1 min directly in the path of a 2-W 488-nm argon laser (protect your eyes!). Alternatively, the FDG solution can be exposed to daylight until the green color is gone. Once bleached, the FDG solution can be freeze-thawed again without rebleaching.

Scaling up
The FACS-Gal reaction can be scaled up without problems. It is important to use an equal volume of cell suspension and 2 mM FDG solution to achieve the osmotic shock at a given concentration of FDG. Finally, add at least 10x volume of ice-cold medium. This step stops the passive osmotic loading by restoring isoosmotic conditions and „freezes" the cell membranes, trapping the substrate and its products inside the cell.

Inhibition of β-galactosidase activity by PETG
With its thiol linkage PETG is nonhydrolyzable by β-galactosidase and thus remains at constant concentrations. The hydrophobic drug readily crosses the cell membrane and can be used to stop the conversion of FDG by adding it at any step to a final concentration of 1 mM.

References

1. Nolan et al.: Fluorescence-activated cell analysis and sorting of viable mammalian cells based on β D-galactosidase activity after transduction of E.coli *lacZ*. PNAS VOL 85, pp2603-2607 (1988)
2. Fiering et al.: Improved FACS-gal: flow cytometric analysis and sorting of viable eukaryotic cells expressing reportergene constructs. Cytometry 12:291-301 (1991)
3. Hofmann and Sernetz: A kinetic study on the enzymatic hydrolysis of fluoresceindiacetate and fluorescein-di-β-D-galactoside. Analyt. Biochem. 131, 180-186
4. Karttunen and Shastri: Measurement of ligand-induced activation in single viable T cells using the lacZ reportergene. PNAS Vol.88 pp3972-3976, (1991)
5. Yancopoulos et al.: A novel fluorescence-based system for assaying and separating live cells according to VDJ recombinase activity. Mol Cell Biol 10:1697-1704, (1990)
6. Fiering et al.: Single cell assay of a transcription factor reveals a threshold in transcription activated by signals emanating from the T-cell antigen receptor. Genes Dev. 4:1823-1834 (1990)
7. Kerr and Herzenberg: Gene-search viruses and FACS-gal permit the detection, isolation and characterization of mammalian cells with in situ fusions between cellular genes and *Escherichia coli LacZ*. Methods, Vol.2, No. 3 261-271, (1991)
8. Reddy et al.: Fluorescence-activated sorting of totipotent embryonic stem cells expressing developmentally regulated *lacZ* fusion genes. PNAS VOL 89, pp6721-6725 (1992)
9. Krall and Braun: In situ lacZ Retrovirus-Marked Lymphocytes Define a B Cell Microenvironment in the Lymph Node Medulla. The New Biologist VOL. 4, pp581-590 (1992)

13 Ligand Acidification by Nonadherent Cells

R.F. Murphy

13.1 Background

The goal of methods for ligand acidification by nonadherent cells is to obtain nearly continuous measurements of the pH to which a receptor-bound ligand is exposed after endocytosis. The protocols presented below are based on the method developed by Sipe and Murphy [1] for adherent (BALB/c 3T3) cells and modified by Sipe et al. [2] for nonadherent (K562) cells. Reviews of acidification of endocytic compartments [3,4] and flow cytometric methods for analysis of endocytosis [5,6] may be consulted for additional background. The protocols below describe analysis of transferrin (Tf) acidification, but they may be easily modified for use with a different ligand.

13.2 Material

Fluorescent transferrin conjugates The ligand to be used must be conjugated with a pH sensitive fluorescent probe, normally fluorescein isothiocyanate (FITC), and, separately, with a pH-insensitive probe, such as lissamine rhodamine sulfonyl chloride (LRSC) or Cy5. Such conjugates may be purchased (Molecular Probes, Eugene, OR) or prepared following standard procedures (see [1] for preparation of Tf conjugates). Conjugates with stable, small molecular weight dyes are preferable since some larger probes (e.g., phycobiliproteins) may be unstable or undergo changes in properties after internalization and acidification in endosomes and lysosomes. Best results are usually obtained with a dual-laser flow cytometer equipped with an argon laser to excite FITC at 488 nm (530-nm emission filter with 30-nm bandwidth) and a krypton laser or dye laser to excite LRSC at 568 nm (625-nm emission filter with 35-nm bandwidth) or a krypton laser or He-Ne laser to excite Cy5 at 647 nm or 633 nm. Stock solutions of Tf conjugates at 1 mg/ml are normally prepared. Each conjugate should be tested for specificity of binding by competition with unlabeled ligand and an optimal labeling concentration chosen. For Tf, this should be 5–10 µg/ml. For the protocols below, a 10 mg/ml stock solution of unconjugated diferric human Tf in phosphate-buffered saline (PBS) is also needed to confirm the specificity of the fluorescent conjugates.

Standard pH buffers are used for calibrating the pH dependence of the fluorescence of ligand conjugates. All contain 50 mM hydroxyethyl-piperazine ethanesulfonic acid, 50 mM 2-[N-morpholino] ethane sulfonic acid (MES), 50 mM NaCl, 30 mM NH$_4$Ac, and 40 mM NaN$_3$. The pH calibration curve requires (at a minimum) buffers adjusted to pH 4.00, 4.25, 4.75, 5.00, 6.00, 6.50, 7.00, 7.25, 7.50, and 8.00 (all when measured at 0°C).

- PBS containing 8 mM NaH$_2$PO$_4$, 2.7 mM KCl, 140 mM NaCl, and 1.5 mM KH$_2$PO$_4$ (adjusted to pH 7.4) is used for washing cells. The appropriate base salt solution for a given cell type may be substituted (e.g., α-minimum essential medium salts).
- RPMI (or other growth medium; without serum) is used for cell labeling.

13.3 Method

Measurement of acidification kinetics for internal ligand requires completion of protocols B, C, and D below. The common labeling procedure, which is the starting point for each of these protocols, is described in protocol A.

A. Cell labeling

1. Collect cells and wash twice with PBS at room temperature.
2. Resuspend cells in a small volume of RPMI (without serum).
3. Count cells.
4. Dilute cells to 10^7/ml with RPMI and place on ice.
5. The cells to be used for one of the protocols should be labeled together. Place the required number of cells in an Eppendorf or other conical tube on ice. Determine the volume(s) of labeled Tf(s) and the additional volume of RPMI needed for a final concentration of 5 x 10^6 cells/ml and a final concentration of 5–10 µg/ml total labeled Tf. When using a mixture of FITC Tf and LRSC Tf, use an equimolar amount of each. When using a mixture of FITC Tf and Cy5 Tf, use a 9:1 ratio (nine times more FITC than Cy5). Add the RPMI to the cells and then add the labeled Tfs. Incubate on ice for 1 h.
6. Prepare parallel „cells only“ (no labeled Tf added) and/or blocked [1 mg/ml unlabeled Tf in addition to the labeled Tf(s)] control samples as called for in the specific protocol.
7. At the end of the labeling time, aliquot into separate samples (if specified in the protocol) and centrifuge to pellet the cells. Aspirate all supernatant, resuspend pellet in full tube of PBS and pellet again. Remove all supernatant and resuspend at the specified concentration in the specified buffer (either PBS or one of the standard pH buffers) as called for in the specific protocol.

B. On-line acidification kinetics

1. Label 10^7 cells with a mixture of FITC Tf and LRSC Tf (Cy5 Tf may be used in place of LRSC Tf if desired) according to protocol A, resuspending the cells after labeling in 2 ml PBS (5×10^6 cells/ml). Split into two samples in flow tubes.

2. Place the first sample on the flow cytometer with ice or ice water in the sample cooling chamber. Acquire list mode data for 5 min to establish starting fluorescence values.

3. Rapidly shift the sample temperature to 37^0C at the same time as beginning acquisition of a new list mode file. Acquire kinetics information either using time as a parameter or recording the number of cells measured during discrete time intervals (ACQ8 method). Accuracy of time recording should be at least 1 s (60 tics when using ACQ8).

4. Continue acquisition for approximately 20 min after-warm up and then repeat from step 2 for the duplicate sample.

5. Analyze the data to obtain mean fluorescence for both FITC and LRSC for live cells (use scatter gate) versus time (use **KINPRO** if using Consort/VAX) (see Fig. 1, panels B,C). Calculate and display ratio of FITC to LRSC versus time to observe acidification kinetics (panel D). If subpopulations are suspected in the cells (not expected for cultured cell lines), display the ratio for each cell versus time using a dot plot or contour map.

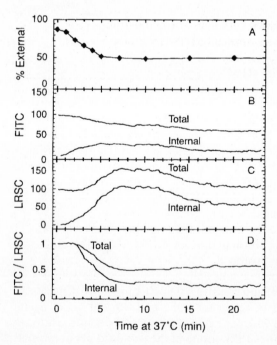

Figure 1. Continuous Tf acidification measurements in K562 cells. FITC Tf fluorescence **(B)** and LRSC Tf fluorescence **(C)** were measured for individual cells as a function of time after warming from $0°C$ to $37°C$. Mean values for live cells (gated using light scatter) were calculated for cells over 10-s intervals. These values were normalized by division with values before warming. Average results are shown for nine experiments. In parallel, Tf internalization kinetics were measured **(A)** using a polyclonal FITC-conjugated antibody to detect surface-bound Tf (average of four experiments). These data were used to subtract the fluorescence of Tf remaining on the surface from the total fluorescence to yield internal fluorescence **(B,C)**. The ratio of FITC Tf to LRSC Tf fluorescence is shown for both total and internal Tf **(D)**. (From [2]).

C. pH calibration

1. Label 6×10^6 cells with a mixture of FITC Tf and LRSC Tf (Cy5 Tf may be used in place of LRSC Tf if desired) according to protocol A. Use the same conjugates as in protocol B. Divide the cells after labeling into 12 aliquots (5×10^5 cells/aliquot) and resuspend each aliquot in 0.5 ml of a different standard pH buffer (use duplicate samples of pH 7.00). Perform the washes in sets (as many as can fit in the centrifuge at one time) and analyze them immediately after resupending in the final buffer (the Tf falls off its receptor with time). Six samples at a time is typical.

2. Repeat step 1 above except do not add labeled Tf to the cells („cells only" controls for each pH buffer).

3. Label 1×10^6 cells in the presence of 1 mg/ml unlabeled Tf (blocked controls), split into two aliquots after labeling, resuspend in pH 7.00 buffer after labeling, and analyze.

4. Calculate mean fluorescence for FITC and LRSC with intact cells (the cells are not necessarily live due to the effects of the pH buffers, so use a forward scatter gate just above the debris) for each sample. Display mean fluorescence versus pH for cells with and without label (use COTFIT if using Consort/VAX). Subtract these two curves (use BIOCAL if using Consort/VAX) to yield corrected fluorescence values versus pH (Fig. 2, panel A). Calculate and display the ratio of these values versus pH (panel B). Note that the conjugation of a pH-insensitive fluorescent probe such as LRSC to a pH-sensitive molecule such as Tf can result in a pH-sensitive fluorescent conjugate (panel A). This need not present a problem with pH calibration as long as the fluorescence ratio is a monotonic function of pH (panel B).

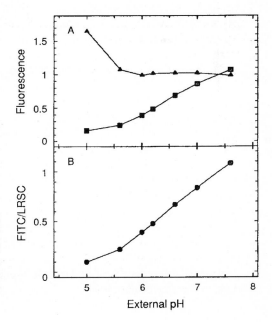

Figure 2. Calibration of pH dependence of fluorescence for receptor-bound Tf conjugates. K562 cells were labeled with both FITC Tf and LRSC Tf, suspended in buffers of different pH and analyzed by flow cytometry at 0°C. Mean fluorescence values were calculated for 10000–20000 cells, and autofluorescence was subtracted before further calculations. Both FITC Tf fluorescence (squares) and LRSC Tf fluorescence (triangles) show pH dependent fluorescence **(A)**, with a marked increase in the LRSC Tf signal observed below pH 5.6. The ratio of FITC to LRSC fluorescence **(B)** decreases monotonically, although both conjugates show pH dependence. From [2].

5. Calculate corrected fluorescence values for the blocked controls and calculate the percentage specificity by averaging the blocked controls and dividing by the average of the unblocked pH 7.00 samples.

D. Internalization kinetics at discrete time points

1. Prepare centrifuge tubes nearly filled with PBS and marked 0, 0.5, 1, 2, 4, 6, 8, 10, 12, 14, and 20 min. Place on ice.
2. Label 7.5 x 10^6 cells with only LRSC Tf or Cy5 Tf (i.e., leave out the FITC Tf) according to protocol A. Resuspend the cells after labeling in 1.5 ml PBS (5 x 10^6 cells/ml).
3. Set up a simulated flow sample chamber (e.g., an outer sample tube attached to a ring stand) to duplicate temperature conditions on the flow cytometer. Start with ice water in the outer chamber as in protocol B. Remove 100 µl cells and put in 0-min tube on ice. Start stopwatch and start 37°C warm-up at the same time. Remove 100 ml samples at above time points and place in appropriate tubes on ice.
4. Spin down cells when convenient, wash once with PBS.
5. Follow either of two options. (a) Label on ice for 30 min with FITC anti-human Tf antibody (to detect surface Tf). Wash twice with PBS and analyze. (b) Resuspend in pH 4 buffer (to strip surface Tf of iron) and immediately centrifuge. Aspirate fully, resuspend in PBS (Tf now dissociates), spin, resuspend in PBS containing 50 mM methylamine (to neutralize internal compartments) and analyze. If using the latter option, keep extra 0-min aliquots and analyze them without stripping to determine the total surface labeling before warm-up.
6. Prepare and flow „cells only" controls also.
7. Calculate mean FITC (if using 5a) or LRSC (if using 5b) fluorescence versus time, subtract value for „cells only", and express as percentage of total surface labeling (Fig. 1, panel A). If using 5b, convert percentage internal to percentage external.
8. Other possibilities for step 5 include (c) protease stripping on ice and (d) crystal violet quenching of external label [7].

Calculation of acidification kinetics for internal ligand

1. Assemble data files containing mean FITC Tf fluorescence versus time [F(t)], mean LRSC (or Cy5) Tf fluorescence versus time [L(t)], and mean percentage of external Tf versus time [E(t)]. Normalize each to 100% at time 0 (start of warm-up).
2. Interpolate E(t) so that it contains values at the same time points as measured in F(t) and L(t).
3. Subtract E(t) point-by-point from F(t) and L(t) to obtain internal FITC Tf fluorescence versus time ($F_i(t)$; Fig. 1, panel B) and internal LRSC Tf fluorescence versus time ($L_i(t)$; Fig. 1, panel C), respectively.
4. For each time point, calculate the ratio of $F_i(t)$ to $L_i(t)$ (Fig. 1, panel D) and convert it to pH internal using the calibration curve generated using protocol C (the curve for K562 cells in Figure 3 was generated using the data in the preceding two).

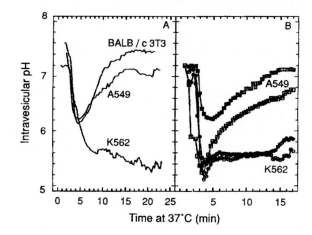

Figure 3. Tf acidification kinetics in different cell types. **(A)** The kinetics of acidification of Tf by K562 cells were calculated from the data in the preceding two figures. For comparison, results for BALB/c 3T3 fibroblasts [1] and A549 epidermoid carcinoma cells [8] are also shown. Note that K562 cells do not alkalinize Tf after acidification, and that the minimum pH (5.4) is significantly lower than that for the other cell types. **(B)** K562 cells were preincubated with 10 μM ouabain (an inhibitor of the Na^+/K^+ ATPase) for 4 h before acidification kinetics were determined (open circles). Results are also shown for control experiments performed on the same day without ouabain (filled circles). Results (from [8]) for A549 cells in the presence (open squares) and absence (filled squares) of ouabain are also shown. Note that ouabain treatment of A549 cells causes acidification to a similar minimum pH as that observed in K562 cells. (From [2]).

References

1. Sipe, D.M., Murphy, R.F. (1987) Proc. Natl. Acad. Sci. U.S.A. 84, 7119-7123
2. Sipe, D.M., Jesurum, A., Murphy, R.F. (1991) J. Biol. Chem. 266, 3469-3474.
3. Mellman, I., Fuchs, R., Helenius, A. (1986) Annu. Rev. Biochem. 55, 663-700
4. Murphy, R.F. (1988) Adv. Cell Biol. 2, 159-180
5. Murphy, R.F., Roederer, M., Sipe, D.M., Cain, C.C., Bowser, R. (1989) Flow Cytometry: Advanced Research and Clinical Applications (Yen, A., ed) Vol. II, pp. 221-254, CRC Press, Boca Raton, Florida
6. Wilson, R.B., Murphy, R.F. (1989) Methods Cell Biol. 31, 293-317
7. Ma, J., Chapman, G.V., Chen, S.-L., Penny, R., Breit, S.N. (1987) J. Immunol. Meth. 104, 195-200 8.Cain, C.C., Sipe, D.M., Murphy, R.F. (1989) Proc. Natl. Acad. Sci. U.S.A. 86, 544-548
8. Cain, C.C., Sipe, D.M., Murphy, R.F. (1989) Proc. Natl. Acad. Sci. U.S.A. 86, 544-548

Part V Cell Sorting

14 Powerful Preselection

C. ESSER

14.1 Background

Fluorescence-activated (FACS) and magnetic cell sorting (MACS) are powerful and sophisticated methods for efficient purification of cells according to specific immunological marker molecules. In certain cases, however, FACS and MACS can be complemented or even substituted by simpler methods of cell separation which make use of either physical differences between cell types, such as density (e.g. Ficoll gradients), biochemical differences, such as the presence of selectable enzymes [e.g. lysosomal enzymes–Leucine methyl ester (LME) lysis], or immunological differences (e.g. „panning" on antibody-coated dishes or complement-mediated lysis). In general, such methods are tailored for specific applications, where they may be of similar efficiency as MACS and FACS, even advantageous for reasons of economy or time, but where loss of positive cells and compromise in the purity of separation is acceptable. Here we describe the methods that can be used to complement FACS and MACS by removing dead cells, erythrocytes, monocytes, and natural killer cells or by enriching rare cells for flow analysis and sorting. A simple method is also provided to destain cells after separation.

The elimination of dead cells and debris, which disturb staining and sorting of live cells, prior to sorting or fixation for flow analysis is an important technical detail because dead cells can spoil the whole experiment. Dead cells are often unspecifically stained or show increased autofluorescence, which is easy to discriminate from cell surface staining by eye but is indistinguishable for the flow cytometer, especially after fixation of the cells for analysis. Dead cells release DNA, thus supporting formation of cell aggregates and clogging the tubes of the flow cytometer.

Preenrichment of cells for flow analysis and sorting can be electronical or physical. In either case it is a manipulation that must be considered critical (see Chap. 3), because it may entail unpredicted effects, especially in pathological situations. For electronic gating, additional parameters with good discrimination are required, and time is still a limiting factor since flow analysis, as a serial method, has a maximum flow rate. Nevertheless, gating out erythrocytes by increasing the forward scatter threshold and thus blindfolding the cytometer for small particles, is a frequently used trick in analysis. Physical preenrichment is advisable if cells could disturb analysis by passive uptake of stain, or if cells that are not easy to gate out electronically are present–especially if they are present in vast excess and due to variation in staining (coefficient of variation) overlap with the population to be analyzed or sorted. For analysis, preenrichment also provides at least one additional optical parameter which would otherwise have been used for gating. For sorting of rare cells, physical preenrichment is in any case advantageous, not only because it increases discrimination, but also because it reduces sorting time and thus stress for cells and budget, and increases the efficiency of sorting due to clearer sorting windows and fewer sort aborts. An ideal method for preenrichment of rare cells for FACS is the use of MACS (Chaps. 15 and 17). Other methods are described here, including panning, rosetting, and LME lysis.

In principle, such methods may be used instead of MACS and FACS for applications where high purity is not required. One should, however, keep in mind that most of the methods work efficiently only for depletion of cells (negative selection), either lysing the cells positive for the selection marker or attaching positive cells to plastic surfaces by multiple binding sites–thus inevitably cross-linking the marker with all physiological consequences and making it difficult to detach the cells from the plastic. MACS and FACS usually provide higher quality for negative selection and allow easy-to-control and efficient positive selection.

14.2 Cotton Wool Columns

The use of cotton wool columns is rather a simple laboratory trick than a full method. However, it is extremely useful when handling cell suspensions prepared from tissues, for example, spleen cells. The homogenized tissue is poured over a cotton wool column and equilibrated with cell buffer which holds back the debris and many dead cells but not live cells. The resulting cell suspension is essentially free of connective tissue and other remnants of homogenization that would only disturb the subsequent staining.

14.2.1 Material

- Pure cotton wool
- Plastic syringes
- Nylon gauze, mesh size greater than 80 µm, cut to squares to fit the diameter of the syringe (e.g., Erbslöh, Düsseldorf, or Schweizerische Seidengazefabrik, Zürich)
- Knife or scalpel

14.2.2 Method

The bottom of a 5- or 10-ml syringe is cut off with a knife (which cuts better if heated). The rim of the syringe is lightly melted on the flame of a Bunsen burner and immediately pressed onto the precut nylon gauze squares yielding a column closed at the bottom with the nylon mesh. The column is filled with approximately 2 ml pure cotton, not tightly stuffed. The columns can be autoclaved for use in sterile tissue culturing. For use, a column is hung into a suitable tube and washed with 5 ml phosphate-buffered saline (PBS) or tissue medium. The cell suspension is taken up in a small volume of medium and added onto the column in the tube. The cells are washed through the cotton wool with 10–20 ml PBS until no further cells are eluted. The column is discarded, and the cells are spun down and taken up in the desired medium and volume.

Note: It is best to lay the nylon gauze on aluminum foil or to work on a steel surface (for example in a laminar flow to avoid burns/stains on a wooden or plastic surface).

14.3 Lysis of Erythrocytes

The lysis of erythrocytes takes advantage of the different susceptibility of enucleated mammal erythrocytes and nucleated leucocytes to hypotonic shock. Erythrocytes are more susceptable than leucocytes and burst rapidly in a hypotonic buffer. The protocol described below has been used primarily with rodent erythrocytes, but it also works with human erythrocytes (which are usually removed by ficollation, see below).

14.3.1 Material

– Tris/NH$_4$Cl solution:
  9 parts 0.83% NH$_4$Cl
  1 part 2.06% Tris(hydroxymethyl)aminomethane
  Adjustment to pH 7.2 with HCl, sterilized by filtration, stored at 4°C, used at room temperature

14.3.2 Method

1. Spin down cell suspension containing erythrocytes.
2. Add 1 ml Tris/NH$_4$Cl solution for 10^8 cells.
3. Resuspend cells and incubate for 5 min at room temperature.

4. Wash cells by filling up with excess volume of PBS or other isotonic medium and spinning down.
5. Repeat if all erythrocytes were not lysed.

14.4 Ficoll Gradient Centrifugation

Ficoll is a nonionic sucrose polymer. Depending on the density properties of certain cell types, they can pass through a Ficoll gradient (e.g. dead cells, erythrocytes), be caught in it (e.g. granulocytes), or remain on top of it in the interphase of Ficoll and medium overlay (e.g. lymphocytes).

14.4.1 Material

– Ficoll solution for removal of dead cells (rodent):
   70 ml 14% (w/v) Ficoll 400 (Pharmacia, Sweden) in H_2O
   30 ml sodium metrizoate (Sigma)
   Mixed, sterilized by filtration, stored protected from light ($\rho=1.36$)
– Ficoll solution for isolation of leukocytes from whole blood (human):
   5.7% (w/v) Ficoll 400
   9.0% (w/v) sodium diatrizoate (Sigma)
   Mix, sterilize by filtration, store protected from light ($\rho=1.077$)
– PBS, tubes

14.4.2 Method

Removal of dead cells
1. Determine absolute number of living cells in the suspension by trypan blue exclusion.
2. Fill Ficoll solution into tubes:
   5 ml in 15 ml tube–holds up to 2×10^7 cells.
   10 ml into 50 ml tube–holds up to 2×10^8 cells.
   The actual volume of Ficoll is not as important as the diameter of the surface in the tube. The numbers given refer to the number of living cells that can be collected from the interphase. Overloading the gradient results in dead cells remaining among the live cells in the interphase. Clear plastic is better than opaque for possible visualization of the interphase.
3. Spin down cell suspension and take up in small volume: 1 ml for gradient in 15-ml tube, 5 ml for gradient in 50-ml tube.
4. Layer cell suspension on the Ficoll cushion without disturbing the Ficoll surface.

5. Centrifuge at 700 g for 15 min (without using brake) at 20°C.
6. Collect living cells from interphase. In my experience, a Pasteur pipette used as „vacuum cleaner" to suck up cells from the top of the Ficoll works best.
7. Dilute collected cells with PBS and wash twice (120 g, 10 min). Dilution is necessary because some Ficoll is always taken along, and cells in Ficoll would require higher centrifugal forces to be spun down.

1. Fill tubes with Ficoll solution (20 ml in 50-ml tube for up to 30 ml diluted blood). **Isolation of peripheral blood mononuclear cells**
2. Dilute blood 1:2 with saline or PBS.
3. Layer blood on top of Ficoll cushion.
4. Centrifuge at 600 g, 30 min at 20^{0}C.
5. Erythrocytes aggregate and pass through the Ficoll to the bottom of the tube; granulocytes enter the Ficoll and remain there.
6. Collect lymphocytes, monocytes, and thrombocytes from the interphase.
7. Wash with cold PBS (120 g for 10 min) until supernatant is clear. The washings remove thrombocytes, which are not spun down at this centrifugal force.

Note: Removal of dead cells is also useful in tissue culturing; however, ficollation is stressful for cells and should not be considered as a routine measurement.

14.5 LME Killing of Monocytes

LME (L-Leucine methyl ester) is a lysosomotropic agent and kills monocytes. This is a good example of the exploitation of a chemical characteristic of a cell to eliminate it specifically from a population.

14.5.1 Material

– 10 mM LME (Sigma) in RPMI 1640 (no FCS)
– RPMI 1640 with and without FCS supplement
– Nylon gauze, mesh size greater than 80 μm, tubes

14.5.2 Method

1. Ficollate and wash whole blood (see above).
2. Take up mononuclear cells from the ficoll gradient interphase at 10^7/ml in RPMI 1640 <u>without</u> FCS.
3. Add one equal volume of LME stock.

4. Incubate at room temperature for 45 min; shake three or four times during this period.
5. Wash twice with RPMI 1640 with FCS (120 g, 10 min).
6. Filter through nylon gauze (see method 1 for the preparation of a column closed at the bottom with nylon gauze).

14.6 „Panning"

Cells bearing substantial amounts of one antigen on the surface (e.g. B cells) can be purified by binding to antibodies coated onto plastic. The bound cells cannot be removed from the plastic dish without loss, but they can be cultivated in the dishes if desired. Enrichment by panning is efficient, but only cells with high levels of antigen expression can be selected. Contamination with cells sticking unspecifically to the plastic dish is another problem restricting the usefulness of this method. However, because only minimal equipment is needed for reasonable results, panning offers an alternative to more sophisticated sorting methods.

14.6.1 Material

– Tissue qualitiy plastic ware (petri dishes or bottles)
– Antibody solution 2 µg/ml (do not filter-sterilize this diluted solution)
– PBS/1% bovine serum albumin (BSA)
– Rubber scraper
– Cells, suspended in medium or PBS/1% BSA

14.6.2 Method

1. Cover plastic dish with antibody solution and incubate overnight at 4 °C.
2. Block plastic dish for 1 h with PBS/1% BSA; discard PBS/BSA.
3. Add cell suspension. On a 15-cm diameter petri dish about 2 x 10^7 cells can be bound; thus take up cells in a volume suitable to cover the plastic dish (e.g. 15 ml for 15 cm Ø dish).
4. Incubate for 30 min at 4 °C; move dish lightly once during this period.
5. Wash unbound cells with ice-cold (!) PBS/1% BSA. This is done by pipetting PBS/1% BSA over the cells and tipping the dish to the side so that the solution can be withdrawn again. Repeat washing until no further cells appear in the washing solution.
6. To use unbound cells, spin down and take up at desired concentration. To use bound cells, scrape them off with a rubber scraper or add culture medium directly onto the cells.

Note: Cells may not bind to the plastic due to several reasons:
- Charge of plastic ware was bad, so that coating did not work.
- Filter sterilization of very low concentrated antibody solutions may result in loss of the protein in the filter.
- Washing solution was not cold, and bound cells were washed off.

14.7 Elution of Cytophilic Antibodies and Antibodies Bound to Cells After Staining

This is not a technique that separates different cells but a trick that one may need to „destain" cells (for example, if it is necessary to stain with a different antibody of the same „color"), or if it must be excluded that the staining reagent detects antigen bound to cells via, for instance, Fc receptors, rather than having been truly synthesized by cells. For instance, cytophilic uptake of antibodies is a major problem in macrophage-containing cell populations. However, the noncovalent antigen-antibody or FcR antibody binding can be disrupted at low pH even if affinities of more than 10^{-11} are involved.

14.7.1 Material

- Acid elution buffer:
  0.05 M sodium acetate/acetic acid, pH 3.5
  0.085 M NaCl
  0.005 M KCl
  1% FCS, sterilized by filtration, stored at 4°C
- PBS/1% BSA or other isotonic, protein-containing buffer

14.7.2 Method

1. Pellet 10^6–10^8 cells in 15-ml tube.
2. Resuspend cells in 1 ml acid elution buffer.
3. Leave at room temperature for 3 min, not much longer, since the number of dead cells increases rapidly.
4. Neutralize by filling up tube with PBS/1% BSA; the pH should be physiological again.
5. Wash cells twice; viability should be more than 90%.
6. Stain cells according to the protocol.

Note: Different cells may be differently sensitive to low pH. If too many cells die in the procedure, reduce incubation time or increase pH to 4–4.5. However, one must be certain that all antibodies are eluted under these conditions.

References

- Boyle, W. (1964) An extension of the 51Cr release assay for the estimation of mouse cytotoxins. Transplantation 6:761
- Bøyum, A. (1968) Isolation of mononuclear cells and granulocytes from human blood. Scand.J.Clin.Lab.Invest. 21:77
- Thiele, D.L., Kurosaka, M., Lipsky, P.E. (1983) Phenotype of the accessory cell necessary for mitogen-stimulated T and B cell responses in human peripheral blood: delineation by its sensitivity to the lysosomotropic agent, l-leucine methyl ester. J. Immunology 131:2282
- Mage, M.G., McHugh, L.L., Rothstein, T.L. (1977) Mouse lymphocytes with and without surface immunoglobulin. Preparative scale separation in polystyrene tissue culture dishes coated with specifically purified anti-immunoglobulin. J. Immunol. Meth. 15:47
- Wysocki, L.J., Sato, V.L. (1978) „Panning" for lymphocytes: a method for cell selection. Proc. Natl. Acad. Sci. 75:2844
- Ishizaka, T., Ishizaka, K. (1974) Mechanisms of passive sensitisation. IV. Dissociation of IgE molecules from basophil receptors at acid pH. J. Immunol. 112:1078
- Kumagai, K., Abo, T., Sekizawa, T., Sasaki, M. (1975) Studies of surface immunoglobulins on human B-lymphocytes. J. Immunol. 115:982

15 High Gradient Magnetic Cell Sorting

S. Miltenyi and E. Pflüger

15.1 Background

Magnetic cell sorting is a powerful tool for the isolation of certain cell subsets from cell populations. So far magnetic cell sorting has been used for isolation of many cell types in human [1] and animals, but also for separation of plant cells and bacteria [2].

Cells themselves show no magnetic properties, except for the very small diamagnetism due to the water they contain. By introducing a magnetic label specific for certain cell types, complex cell mixtures can be fractionated using magnetic forces. The absence of a natural magnetic background in cells permits the use of very small colloidal magnetic particles (diameter < 100 nm) for specific magnetic labeling. However, in this case high gradient magnetic techniques producing extremely high magnetic forces have to be used for separation. Small particles have several advantages over large particles for magnetic labeling and efficient separation of cells [3].

Cells labeled with magnetic particles can be simultaneously stained with fluorochromated antibodies to control and evaluate the quality of magnetic separation by flow cytometry or microscopy. Since in many respects the physical behavior of colloidal particles is similar to that of macromolecules, like fluorochromated antibodies, labeling with superparamagnetic colloidal particles can be termed a „magnetic staining".

In this chapter, we will discuss separation techniques based on staining with colloidal superparamagnetic particles and separation of stained cells using high gradient magnetic fields (*MACS* cell sorting system [3], Figures 1 and 2).

A single cell suspension of sufficient quality is essential for a good magnetic separation. Methods for obtaining such single cell suspensions are the same as those for flow sorting (see Chap. 14). Since the magnetic separation process cannot distinguish between single cells and aggregates of cells, it is very important to avoid cell aggregates which might contain mixtures of „wrong" and „right" cells. The cells should be re-suspended carefully (e.g., by finger flipping the tip of the tube) especially after centrifugation. Protein or EDTA may be added to the medium to reduce cell aggregation during handling.

Preparation of cells

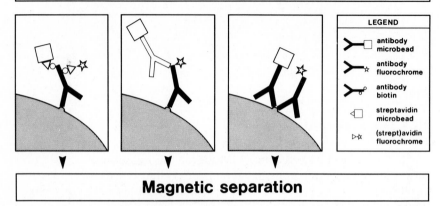

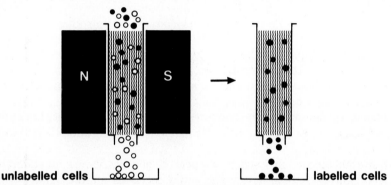

Figure 1. Methods for simultaneous magnetic and fluorescent staining of cells. *Left:* labeling method using avidin-biotin system. *Middle:* labeling with primary, (fluorochromated) antibody and second step microbeads. *Right:* labeling using direct magnetic and fluorescent conjugates

Depletion or enrichment The optimal strategy for any specific cell separation depends on
– The availability of suitable cell surface markers.
– The characteristics of the original cell mixture (frequency of wanted cells, characteristics of the other cells).
– Specific requirements with regard to purity of sorted cells and activation status of cells which might be affected by antibody staining.

Positive isolation requires a cell surface marker which is specific for the cells of interest. Since the separation process distinguishes only between nonmagnetic and magnetic cells, it is important that the negative cells really remain unlabeled (no background staining).

Depending on the cell type and the surface marker used for positive cell sorting, the stained and sorted cells might be functionally influenced, due

Figure 2.
MACS cell
separator.

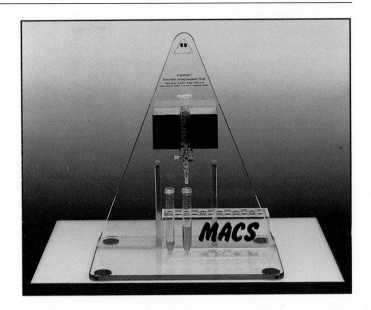

to the labeling with antibodies and microbeads. This is a problem inherent to staining with antibodies which cross link the antigen on the target cell surface and thus may activate or suppress the antigen expressing cell. Labeling with colloidal magnetic particles has no additional effects on the functional status of the stained cell as far as known so far. Beads coupled to the surface of a cell will be capped off or endocytosed and degraded, depending on the cell type, antigen and microbeads used.

Similar beads as the beads used in the MACS System have been used to treat iron deficiency in humans and animals and have been shown to be biodegradable and non toxic [4].

Negative cell sorting, for example, labeling and depleting of cells from complex mixtures, usually will not give homogenous cell populations but the depleted non-stained cells are functionally naive. Depending on the quality of the staining, however, very efficient depletion can be obtained where more than 99,5% of the stained cells will be depleted. A side effect of depletion is that dead cells, debris and cell clumps are often removed from the cell suspension, due to unspecific labeling of these components.

Two cycles of separation, first a depletion, followed by positive enrichment can be combined for efficient isolation of rare cells, or for depletion of unwanted or unspecifically staining cells before a positive separation.

Presorting for fluorescence activated cell sorting

Due to the small size of the magnetic microbeads, magnetic cell sorting with *MACS* is a valuable complementation to flow cytometry and fluorescence-activated cell sorting. Presorting with *MACS* can be used to enrich rare cells, thus cutting down sorting time, allowing efficient sorting grades and flow rates.

Magnetic Labeling Several methods can be used to obtain specific magnetic labeling: For many antigens on human or mouse leukocytes, surface-antigen specific magnetic particles are available commercially. These microbeads are conjugated to surface-antigen specific antibody.

Alternatively, primary antibody used for cell surface labeling can be detected by anti-immunglobulin-specific microbeads or biotinylated primary antibody by streptavidin-conjugated microbeads.

Titers for primary antibodies should be determined carefully using flow cytometry, to avoid background staining (see Chap. 3). Background staining reduces significantly the performance of the separation process (recovery when doing a depletion, purity of the positively enriched cells for enrichment) because even slightly labeled cells will be retained in the magnetized column.

Mixtures of different primary antibodies, detected by secondary beads can be used, however, each antibody has to be titrated.

The speed of reaction of colloidal magnetic microbeads to the cell surface antigen is slow compared to staining with cell surface specific antibodies. However, because high gradient magnetic separation can be made extremely sensitive, magnetic staining does not have to be completed to the saturation point, usually a few minutes of incubation with the beads are sufficient for labeling. Only a few dozen of colloidal particles are needed on a single cell for magnetic separation.

In order to achieve reproducible separations, it is important to use defined bead concentrations, incubation times, and temperature. Unless the cells can be separated immediately after staining, beads should be washed away after magnetic labeling, to prevent magnetic background staining. The beads will not sediment at accelerations typically used to spin down and wash the cells.

Like any staining reagent, also the magnetic beads may be titrated. This can be done functionally, using different concentrations of beads for one particular, standardized separation and determining the concentration with the best enrichment rate, for example, low background and high specific labeling. Since the magnetic labeling reaction is typically in its linear phase the degree of magnetic labeling will depend on the bead concentration, incubation time, but also on density of the antigen on the cells.

Only few cell surface antigens will be stained with colloidal magnetic particles under conditions typically used for magnetic separation. Therefore it is possible to label the remaining antigens with fluorochromated antibody of the same specificity to control the efficiency of separation.

When using a two step method for magnetic labeling either fluorescent second step reagents can be introduced after magnetic staining or a fluorochromated primary antibody can be used.

The colloidal *MACS* magnetic microbeads are too small to be detected by light microscope or flow cytometer. Thus magnetically labeled cells behave the same as unlabeled cells for flow analysis and fluorescence activated cell sorting (see Chap. 17).

Magnetic separation of the labeled cells is carried out using high gradient **Magnetic** separation columns. These columns are filled with thin plastic coated **separation** ferromagnetic fibers, which are magnetized in the field of a strong permanent magnet (**MACS** columns and separator, Miltenyi Biotec).

The extremely strong magnetic forces generated in the neighborhood of the magnetized fibers enable separation of cells labeled with colloidal superparamagnetic microbeads. The probability for a cell being caught in the matrix depends on the strength of magnetic labeling of the cell, the flow rate of the cell suspension through the column and the geometry of the column.

The sensitivity of the separation can be influenced by varying the flow rate of cells through the column. For depletion, a slow flow rate should be chosen, whereas for enrichment it is better to use a faster flow rate combined with multiple passages over the column. As mentioned before, magnetic labeling and optimal flow velocities depend on each other. For individual antibodies it is therefore necessary to optimize separation efficiency by adjusting the speed of flow. If the antigen used for separation is expressed on more than one cell population but in different amounts it is possible to enrich the different populations. The efficiency is dependent on the difference of labeling.

Generally, separations should be checked for purity specifically by flow cytometry or fluorescence microscopy. Like for other separations, the protocol should contain information on the frequency of cell populations before and after magnetic sorting, allowing to calculate the enrichment/depletion rates. Numbers of live and dead cells, allow to calculate the recovery rates. Also, information on the reagents, the size of the column and the flow rate (ml/min.) should be part of the protocol.

Enrichment and depletion rates should be calculated by:

Enrichment rate:

$$f_E = \frac{\% \text{ neg in original sample}}{\% \text{ pos. in original sample}} \times \frac{\% \text{ pos. in pos. fraction}}{\% \text{ neg. in pos. fraction}}$$

The enrichment rate represents the average number of negative cells passing through the column per negative cell trapped unspecifically (assuming that all positive cells were retained). This value is a measure for the absence of unspecific labeling and trapping. Typical values range between 100 to 1 000 for a clean staining and good separation.

Depletion rate:

$$f_D = \frac{\% \text{ pos. in original sample}}{\% \text{ neg in original sample}} \times \frac{\% \text{ neg. in neg. fraction}}{\% \text{ pos. in neg. fraction}}$$

The depletion rate represents the average number of positive cells being trapped per positive cell passing through the column. It is a measure for the strength of the magnetic labeling compared to the flow rate. Typical values range between 100 to several thousand depending on the number of antigenic sites available on the cells, the quality of labeling and the speed of separation.

Separation of Monocytes, B and T Cells from Human Peripheral Blood Mononuclear Cells (PBMC) by *MACS*

15.2 Material

- Magnetic Cell Separator (*MACS*) with separation columns (type A2; B1; B2).
- Flow-cytometer (optional) or fluorescence-microscope.
- Centrifuge.
- Microscope, Neubauer cell counting chamber, trypan blue for vital staining.
- Phosphate-buffered saline (PBS/BSA)(PBS with 0.5 % bovine serum albumin).
- 70 % ethanol in water (to fill the *MACS* column without trapping bubbles).
- Ficoll hypaque (Pharmacia, Uppsala).
- For monocyte separation: *MACS* CD14 Microbeads, CD14-FITC (LeuM3 (Becton Dickinson), fluorescein conjugate 25 mg/ml).
- For B cell separation: *MACS* CD19 Microbeads, CD19-PE (Leu12, phycoerythrin conjugate).
- Autologous platelet free serum (platelets removed by spinning at 5 000 x *g* for 20 min).
- For indirect T cell separation: *MACS* Rat-anti mouse IgG1 microbeads (RamG1).
- CD3-FITC (anti-Leu-4, fluorescein conjugate, 15 mg/ml) (mouse IgG1).
- Propidium iodide (PI)(1 mg/ml in PBS).
- 2×10^8 Peripheral blood mononuclear cells (PBMC) isolated by Ficoll hypaque (see Chap. 14), alternatively buffy coat can be used.

15.3 Method

Setting up a MACS separation column
- Attach 10 ml syringe, filled with ethanol to the side-plug of the stopcock.
- Fill appropriate separation column from the bottom with ethanol (without trapping air-bubbles in the matrix).

- Place column in the **MACS** , close stopcock, and attach a syringe filled with PBS/BSA at the side-plug of the stopcock.
- Wash column with several column volumes of PBS/BSA from the top of the column.
- Before separation, wash the column with ice-cold PBS/BSA to cool the column, then attach flow resistor (needle) at the bottom plug of the stopcock (small diameter of needle: low flow rate, large diameter high flow rate).

15.3.1 Depletion and isolation of monocytes (CD14)

- Fill B2 column as described above.
- Add 200 ml **MACS** CD14 microbeads to 2×10^8 PBMC, resuspended in 800 µl PBS/BSA.
- Incubate for 15 min. at 6° to 12°C (not on ice!).
- Add 80 µl CD14-FITC and incubate for 5 min. on ice.
- Wash once with cold PBS/BSA (10 ml).
- Place column in **MACS** separator.
- Resuspend cell pellet in 2 ml PBS/BSA, take an aliquot of about 10^5 cells for analysis in flow cytometer or fluorescence-microscope.
- Place collecting tube below column.
- Apply the cells on top of a pre-filled B2 column with a 22 G needle attached.
- Let the negative cells pass the column, wash with 15 ml PBS/BSA. Collect as „negative fraction".
- Remove needle, take column out of the **MACS** , attach a 10 ml syringe filled with PBS/BSA via adapter on top of the column and collect positive fraction by flushing column with 15 ml PBS/BSA.
- Count cells of both fractions in Neubauer chamber with trypan blue.
- Add PI (1:1 000 dilution) to aliquots of each fraction(10^5 cells) and analyze purity and viability by flow cytometry or microscope.
- Calculate recovery.
- The negative fraction can be further fractionated using other markers.
- The purity of the positive fraction may be increased by a second **MACS** separation:

15.3.2 Isolation of monocytes(CD14)

- Apply positive cell fraction to A2 column, separate with a 23 G needle, and wash with 3 ml PBS/BSA.
- Cycle positive fraction: Take the column out of the **MACS** , flush the bound cells in the upper reservoir, and put the column back in **MACS** . Let the cells run through again and wash with 3 ml PBS/BSA.
- Repeat cycle step once or twice.

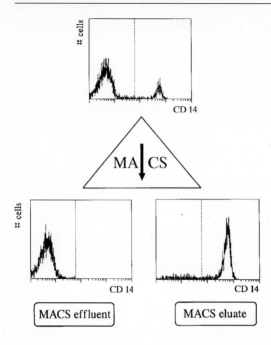

Figure 3. Monocyte separation. PBMC were labeled with CD14 microbeads, stained with CD14-FITC and separated as described Typical FACScan histograms of original, negative and positive fraction

- Take the column out of the separator and flush out the positive fraction.
- Check purity and determine cell number.
- This isolation can be done using PBMC or buffy coat (Fig. 3).

15.3.3 Separation of B cells (CD19)

For this sort the negative fraction of the previous separation, PBMC or buffy coat may be used. Isolation from buffy coat will result in the most pure B cells, predepletion of monocytes is usually not necessary (Fig. 4).
- Resuspend 10^8 cells in 400 µl PBS/BSA, supplemented with 20% platelet free autologous serum.
- Add 100 µl **MACS** CD19 microbeads, incubate for 15 min. at 6° to 12°C.
- Add 50 µl CD19-PE, stain for 5 min at 0° to 4°C and wash once with PBS/BSA.
- Resuspend cell pellet in 0.5 ml PBS/BSA (save aliquot for quality control).
- Apply cells to A2 column, separate with a 24 G needle, and wash with 3 ml PBS/BSA. Collect cells as negative fraction.
- Cycle the positive fraction 2–3 times as described for the monocyte separation using 23G needle.
- Take the column out of the separator and flush out the positive fraction.
- Add PI (1:1 000 dilution) to quality control samples, determine purity, viability, and number of cells (see above).

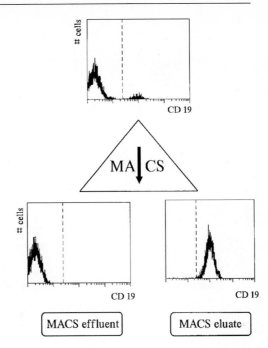

Figure 4. B Cell separation. PBMC were labeled with CD19 microbeads, stained with CD19^{PE} and separated as described. Typical FACScan histograms of original, negative and positive fraction

15.3.4 Separation of T cells (CD3)

For this sort the negative fraction of the CD14-separation or PBMC can be used. The protocol is included to demonstrate indirect labeling using Rat-anti-mouse IgG1 microbeads (Fig.5).

– Resuspend 10^8 cells in 350 µl CD3-FITC.
– Incubate for 10 min at 4° to 8°C.
– Wash with 10 ml PBS/BSA.
– Re-suspend pellet in 400 µl PBS/BSA and add 100 µl **MACS** Rat anti mouse IgG1 microbeads.
– Incubate for 10 min at 6° to 12°C. Add 500 µl PBS/BSA.
– Apply cell suspension on top of B1 column and separate in two cycles (see Chap. 15.3) with a 23G and a 21G needle.
  Washing negative cells out, flushing positive cells back up.
– Collect negative fraction in 15 ml, positive fraction in 10 ml.
– Add PI (1:1 000 dilution) to samples, analyze purity, viability, and loss of cells.

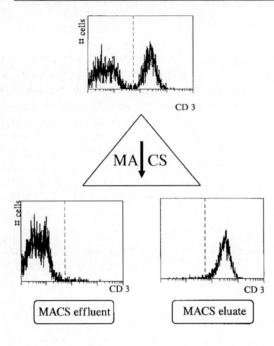

Figure 5. T Cell separation. PBMC were stained with CD3^{FITC}, labeled with Rat anti mouse IgG1 microbeads and separated as described. Typical FACScan histograms of original, negative and positive fraction

15.4 Tips, Tricks, and Troubleshooting

15.4.1 Staining

- Avoid capping of antibody on cell surface during staining. Work fast, keep cells cold, use cold solutions only.
- Working on ice requires increased incubation times for antibody microbeads.
- Increased temperature and prolonged incubation time for staining may result in unspecific cell labeling.
- Increased antibody microbeads concentration may impair antibody fluorochrome staining, when antibody of the same specificity is used.
- Microbeads can be sterilized by filtration through 0.2 µm GV 4 filter (Millipore S. A. Molsheim/France).

15.4.2 Columns

- The capacity of columns applies to bound cells.
- For good enrichment of cells use as small a column as possible. Overloading is advantageous for better enrichment of positive cells.
- Separation columns may be autoclaved at 120°C.
- Flow rates for depletion or enrichment depend on antigen expression on

cell surface. Separation conditions as flow rate, incubation time, and antibody titer for individual antibodies have to be optimized, when second step reagents are used.

– For complete depletion of cells use column with sufficient capacity.
– Do not introduce tissue or big cell clumps to separation column since they may be trapped irreversibly and hence can destroy the column. (see Chap. 14 for removing debris)
– Never let column run dry.
– It is important to avoid air bubbles in the column, use degassed solutions only.

15.4.3 Analysis

– Use PI to gate out dead cells by scatter and PI-fluorescence, (forward scatter(FSC) / side scatter(SSC) and fluorescence 2(F2) / fluorescence 3(F3)) for quality control.

15.4.4 Troubleshooting

1. Bad purity of positively separated cells:
– Use faster flow rate.
– Use lower titer of reagents.
– Add cycle procedure.

2. Poor recovery of positive cells in positive fraction:
– Use slower flow rate.
– Use column with sufficient capacity.
– Remove dead cells by Ficoll hypaque.
– Improve magnetic labeling i.e., use concentration titer of reagents or longer incubation times for beads.

3. Poor recovery of negative cells in negative fraction:
– Use faster flow rate.
– Improve magnetic labeling (background staining) i. e. use lower concentration of reagents or shorter incubation times for beads.

4. Positive cells are lost (neither in positive nor in negative fraction):
– Check that no capping has occurred.
– Check for loss during centrifugation.
– Flush column again.

References

1. Kemshead J. and Ugelstad J.: Magnetic separation techniques: their application to medicine, Molecular and cellular Biochemistry 67, 11-18 (1985)
2. Kronick P. and Gilpin W.: Use of superparamagnetic particles for isolation of cells. J. Biochem.. and Biophys. Methods 12, 73-80, (1986)
3. Miltenyi S., Müller W, Weichel W. and Radbruch A.: High Gradient Cell Separation with MACS, Cytometry 11:231-238 (1990)
4. Muir A. and Goldberg L.: Observations on subcutaneous macrophages. Phagocytosis of Iron Dextran and ferritin synthesis Quart. J. Exp. Physiol. 46,290 (1961)

16 Setup of a Flow Sorter

C. Göttlinger, B. Mechtold, K.L. Meyer, and A. Radbruch

16.1 Background

The option of sorting out cells according to microscopic parameters has made flow cytometry one of the most valuable tools for cell biology. Sorting makes use of either of two different technical principles. One is based on changing the direction of flow of individual cells electromechanically [4]. The cells are pushed in distinct flow channels of a closed flow chamber. The rate of sorting is limited to about 1000 particles per second. Because the particles do not have to pass a narrow nozzle, fragile and large cells or aggregates can be sorted with ease. Since the entire setup is closed, no aerosols are formed, and the risk of infection is minimal. This sorting principle is used in Partec machines and the FACSort of Beckton-Dickinson. Since we have no experience with these instruments we do not discuss them further. The second sorting principle is based on packing the cells into tiny droplets after analysis, charging those droplets that contain cells to be sorted out, and deflecting the charged droplets in an electrical field. This technology is used in free flow-in-air cytometers and allows sorting rates of up to 10^4 particles per second. The arrangement of droplet sorting is shown schematically in Figure 1.

To stabilize the droplet formation and ensure a constant distance between breakoff point and point of laser intersection, a vibration (drop drive) defined by frequency (drop-drive frequency), amplitude, and phase is coupled to the liquid jet via a piezoelectric transducer. A typical value of the drop-drive frequency for a nozzle with an orifice diameter of 70 μm, the jet running at 10 m/s, is 27 kHz, breaking the jet into 27 000 uniform droplets per second at a defined distance from the nozzle tip.

After leaving the nozzle, the cells first pass the laser intersection point, where they are analyzed. The electronic sorting device determines electronically whether the cell fulfils the preset requirements for deflection. The cells travel to the breakoff point in a constant time, the so called "drop delay" time. At the breakoff point, the cells are packed into single droplets. If a droplet is to be deflected, the whole jet is charged for the time of drop formation, then decharged again, leaving the newly formed droplet charged. All droplets then fly through a static electrical field between two metal plates (deflection plates). Charged droplets are deflected from the vertical main stream into a separate collecting tube.

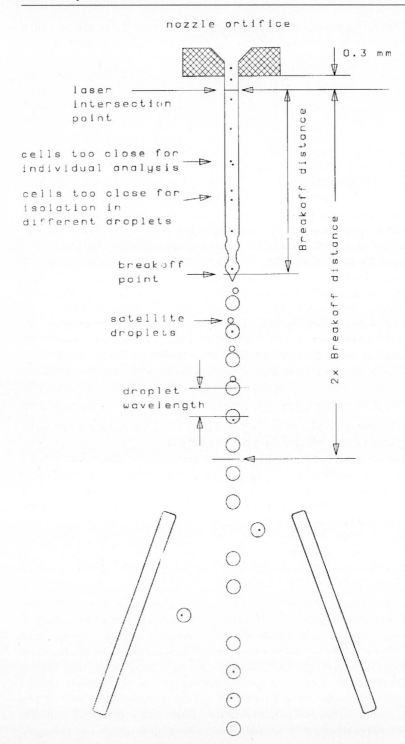

Figure 1. Liquid jet of a sense in air flow cytometer with droplet breakoff

The critical parameter for good sortings is the drop delay, i.e., the time between analysis and droplet formation, which must be constant to allow defined charging of droplets. To account for small irregularities in drop delay and for cell velocity, for good recovery of sorted cells not only the droplet containing the cell but also the droplets before and after can be deflected by merely increasing the time of charging. For the same reasons, better sorting results are obtained with short drop delays, i.e., breakoff points close to the nozzle. For a given diameter of the nozzle ortifice and given viscosity of sheath fluid, the distance between analysis and breakoff points depends on sheath pressure, drop-drive amplitude, and drop-drive frequency. Setting the breakoff point requires optimization of all three parameters. The lowest sheath pressure that still gives a continuous jet and constant drop formation yields the shortest breakoff distance. For the drop-drive frequency, certain frequencies give resonance with nozzle and nozzle holder, self-enforcing the amplitude and stabilizing the position of the breakoff point. Such resonance frequencies can be found easily by observing the breakoff point with the alignment microscope while constantly increasing the frequency from 20 kHz. Resonance frequencies show up because they cause the breakoff distance to shorten. Observing the breakoff point during change of drop-drive frequency, resonances are found with minimum breakoff. Such resonance frequencies should be selected for sorting. Theoretically, the shortest breakoff distance is achieved when the wavelength (droplet spacing) is 4.5 stream diameters [1].

For flow-in-air sorters, sorting speed is limited principally by three factors:

- Cycle time or dead time of the electronics, i.e., time needed for processing data of one cell (typical value: 15–20 µs).
- Time for generating one droplet (reciprocal of drop-drive frequency).
- Statistical distribution of cells. Some cells are always too close together for individual analysis or droplet formation. Such cells are prevented from being sorted by "abort" electronics, lowering the yield of sorting. For a detailed discussion of optimal flow rate at given electronic dead times, drop-drive frequency, and percentage of positive cells see [2]. For high recovery the abort electronics is switched off (ENRICH), lowering the purity of sorted cells. For high purity, abort electronics is switched on, lowering the recovery.

As a basic rule, every sort should be accompanied by a sort protocol containing information on the instrument settings and the purity and recovery of sorted cells (see Table 1).

16.2 Method

1. Alignment and calibration of the instrument for analysis of cells is described in Chap.1. (Use a nozzle with 70-µm orifice for sorting of lymphocytes.)

Setting up a flow-in-air sorter for sorting

Table 1. Protocol of a cell separation

Separation technique	Start population	Fraction	Cell number[a] times 10^6			Purity[c] (%positive)	Recovery[d]
			Living	Dead	Positive[b]		
Centrifugation	Whole blood	Before After				—	—
Lyse and filtration	Peripheral blood mononuclear cells	After				—	—
MACS	Peripheral blood lymphocytes	Positive Negative					
FACS	MACS positive	Deflected					

MACS: Magnetic cell sorting
[a] Haemocytometer or Coulter counter
[b] Frequency of positive cells (Flow cytometer/ Fluorescence microscope)x cell number
[c] Frequency of positive cells
[d] Ratio of number of positive cells recieved from sorting and number of positiv cells in the unsorted sample.

2. Switch fluidics control to SHEATH position; switch on drop-drive frequency and stroboscope for observation of droplets.
3. Set drop-drive frequency to 20 kHz and amplitude to two thirds of maximum.
4. Lower the sheath pressure, observing the breakoff point using the alignment microscope until drop formation is no longer stable; then increase the sheath pressure about 1 psi.
5. Slowly increase the drop drive frequency, while observing droplet breakoff distance. Find a resonance frequency which shortens and stabilizes the breakoff distance (usually around 27 and 33 kHz).
6. Use the drop-drive amplitude to set the breakoff distance somewhere between 2.5 and 4 mm. Avoid setting the amplitude to minimum or maximum since you may have to change it during the sort to adjust the breakoff point.
7. Determine the drop delay: Set horizontal cross-wire of the alignment microscope to the height of the breakoff point. Set the cross-wire to a position twice the distance of breakoff.
8. Move microscope upward, counting the drops (drop plus interim space = 1) until the breakoff point is reached. Use drop number minus 2 for setup of drop delay. Select 1 drop to be deflected.
9. Run sample of chicken red blood cells (CRBCs) and set sorting windows to deflect single CRBCs.
10. Switch on deflection plates and stream charging pulses.
11. Determine the optimal drop delay: Deflect 1000 cells on a slide, increase drop delay for one-half droplet, move slide, and sort again 1000 cells. Repeat until the initial value plus 2 is reached.

12. Determine the deflection spot containing most cells by microscope. Use the corresponding drop delay for sorting.
13. Set the number of deflected droplets to 2 for regular sorts or 1 for sorting of rare cells.
14. Remove CRBC sample and allow backflushing for approximately 1 min. Install sample filter. To sterilize, dip the sample filter and tubing into alcohol, washing off the alcohol with sterile PBS afterwards.
15. Install sample tube with 10^7 cells/ml of the cells to be sorted.
16. During the sort check breakoff distance regularly and correct if necessary by changing the drop drive amplitude or phase, but never the frequency as this would also change the drop delay.
17. Keep cells cool and in suspension during sorting.
18. After sorting determine recovery and purity of sorted cells by microscopy or flow cytometry and do not forget the sort protocol (table). For determination of recovery rate, the absolute cell numbers immediately before and after sorting and the frequencies of sorted cells must be determined.

16.3 Tips, Tricks, and Troubleshooting

There are many ways to sterilize a cell sorter–but most contaminations come with the sample! To keep sterile cell suspensions sterile through sorting it is sufficient to sterilize the compressed air and sheath fluid by fresh anti bacterial filters (0.2 µm pore size). Sample line and nozzle can be sterilized by flushing with 0.1% 7X (Flow laboratories) or any other detergent that is tissue-culture "compatible." In any case the sample line should be backflushed extensively with sterile sheath fluid before running cells. Alcohol, if used for sterilizing the sample line, may agglutinate debris from the wall of the sample line and thus cause clogging. **Sterile sorting**

With sorting for longer than 30 min sedimentation of the cells in the sample tube may become a problem. To prevent sedimentation a "sample medium" of equal density to the cells can be used [3] or the sample tube agitated from time to time. **Long-term sorting**

– Check for air bubbles in the nozzle tip or nozzle holder. Remove sample, switch off deflection plates, and switch fluidics control between OFF, FILL, and SHEATH positions to remove air bubbles. Control for success. If this did not help, the nozzle holder must be removed, to clean the nozzle, with a syringe or ultrasonics. **Unstable breakoff point**
– Check for leakage in the tubings.
– Check for aggregates in the sample; if positive, filter the sample through nylon gaze of 50 µm pore size (e.g., Erbslöh, Düsseldorf, or Schweizerische Seidengazefabrik, Zürich).

– Check for salt crystals at the nozzle tip; remove by gently touching the nozzle tip with a piece of filter paper (e.g. Whatman, Maidstone) while the laser is blocked.

Dispersed side streams
– Control side streams with cold, focused bright light source.
– Use phase control to align the end of liquid stream. The tip of the jet should look like that illustrated in the figure (see above).

Unstable flow rate
– Check for air or liquid leaks of sample tube and tubings, especially the O ring seal between sample tube and sample holder.
– Check sample for aggregates and filter, if necessary, through nylon gaze.
– Check the sample line for clumps and aggregates, especially the connections of tubings and the tubing at the pinch-off valve. Exchange sample line if necessary.

Decharging of collection tubes
Deflected droplets deliver their charge to the collection tube, sometimes causing high local charges of the same polarity as the deflected droplets, causing rejection of later deflected droplets. This can be prevented by grounding the collection vessel. A bent sterile injection needle connected to ground and pinched into the side wall of the collection tube is sufficient for this. Also, it is useful to fill the collection tube with some medium, not only for the comfort of the cells but also for distribution of the charge of deflected drops.

Bad recovery
Determine cell numbers immediately before and after sorting, otherwise the losses due to centrifugation of the cells are added the losses from sorting.

References

1. Kachel V., Fellner-Feldegg H., Menke E. (1990), Hydrodynamic properties of flow cytometry instruments. In: Melamed M.R., Lindmo T., Mendelsohn M.L. (eds) Flow cytometry and sorting. New York: John Wiley & Sons, Inc. pp 27-44.
2. Lindmo et al. (1990), Flow sorters for biological cells. In: Melamed M.R., Lindmo T., Mendelsohn M.L. (eds) Flow cytometry and sorting. New York: John Wiley and Sons, Inc. pp 145-169
3. Mechtold B., Miltenyi S., Irlenbusch S., Göttlinger C., Radbruch A., (1990), Sorting of rare cells. In: Proceedings of the 3rd European Cytometry Users Meeting 1989, Ghent Becton-Dickinson Press, pp74-81
4. Göhde, W. and Dittrich, W. (1971), Impulsefluorometrie – ein neuartiges Durchflussverfahren zur ultraschnellen Mengenbestimmung von Zellinhaltsstoffen. In: Acta Histchem., Suppl. X, p.42–51, 1971.

17 Sorting of Rare Cells

W. WEICHEL, S. IRLENBUSCH, K. KATO, and A. RADBRUCH

17.1 Background

For sorting of cells occurring at frequencies of below 1%, positive selection, i.e., isolation of rare cells on the basis of a molecule that they express, is much more efficient than depletion of the majority of cells, which usually show heterogeneity with respect to any marker that can be used for depletion, making isolation of the rare cells impossible. As one of the most powerful methods for positive selection, fluorescence-activated cell sorting (FACS) has been used for the enrichment and isolation of rare cells ever since its introduction. Examples include immunoglobulin switch variants [1,2], MHC mutants [3,4], and transfectants [5,6]. Such cells occur at frequencies of less than 10^{-2} down to 10^{-8} in a given population. Formerly, rare cells were isolated in several rounds of cell sorting, expanding the enriched cells between the sorts in vitro, thus gradually increasing the frequency of variants until isolation of rare cells and cloning became possible. Under optimal conditions, enrichment rates of up to 10^3 per sort could be achieved, but the number of sorted cells was too low, for reasons indicated below, to allow immediate resorting for purification of the rare cells. Naturally, such experiments could then only be performed with transformed cells, adapted to continuous growth in culture.

Today additional parameters are available in flow cytometry, and new methods for fast and efficient enrichment of cells prior to FACS sorting have been introduced, substantially facilitating the isolation of rare cells.

Basically, the problem with the isolation of rare cells is twofold:

1. Discrimination between rare cells and cells of the major population is difficult because of the variation in immunofluorescence (coefficient of variation) which may lead to a population overlap (Fig. 1), even in otherwise acceptable staining. This and the low absolute number of rare cells makes it difficult to analyze enough cells to identify the population of rare cells with some statistical reliability. This problem is diminished by optimizing the immunofluorescence staining, introducing additional parameters (Chap. 3) and enriching rare cells by fast, parallel magnetic cell sorting (*MACS*) sorting (Figs. 1, 2; Chaps. 14, 15) before FACS analysis and sorting.

2. Large cell numbers must be processed for the sorting of rare cells to obtain enough cells for subsequent analysis. For FACS, cells are

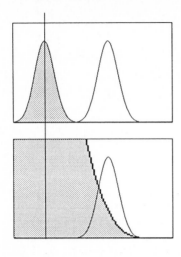

Figure 1. Population overlap. This is one reason why rare cells are difficult to isolate. With regard to any parameter, discrimination between rare and the other cells depends not only on the difference in mean fluorescence but also on the % CV of the major cell population, i.e., the shape of the fluorescence distribution (see Chap. 3), as illustrated here. An overlap of the population of rare cells and the infrequent, diverging cells from the major population is the result, making it impossible to define discriminatory gates for sorting. Population overlap is minimized (a) by optimizing the staining conditions, i.e., maximizing the difference in mean fluorescence, (b) by first enriching the rare cells to equalize population sizes and minimize the effect of variation (e.g., see subsequent figures), and (c) by using as many discriminative parameters as possible, even if they discriminate only between some of the cells of the major population and the rare cells (see Fig. 3)

analyzed one by one (serially) at flow rates of up to 5000 cells per second, i.e., 1.8×10^8 cells in 10 h. This long sorting time means stress for the cells (and the experimenter), and for some applications sorting rare cells out of 2×10^8 cells does not yield sufficient numbers. This problem is solved by use of powerful new parallel enrichment methods which can handle large cell numbers, analyzing and sorting all cells simultaneously. Combination of enrichment of the rare cells by fast, parallel *MACS* (Chap. 15), limited by the use of only one parameter for discrimination, followed by further enrichment with slow, serial FACS, offering several parameters for optimal discrimination, can yield enrichment rates of up to 10^8 fold in one experiment (Fig. 2). Few biological problems require higher enrichment rates.

17.2 Material

The experimental strategy outlined in the figure must be tailored for the individual problem with respect to the frequency of rare cells, markers available for discrimination, and the use of sorted cells. The cell sorting methods to be combined for rare cell sorting are described in detail in Chaps. 14–16. Here we emphasize peculiarities of the sorting of rare cells:
- Single cell suspension of good viability (Ficoll-Hypaque centrifugation removes dead cells and debris. Chap. 14)
- Phosphate buffered saline (PBS) with 1% fetal calf serum (FCS) or bovine serum albumin (BSA) (PBS/FCS or PBS/BSA)
- Propidium iodide (PI), 1 mg/ml in water
- Immunofluorescent and magnetic staining reagents in PBS/FCS with 0.03% sodium azide. For sterile sorting, reagents must either be steri-

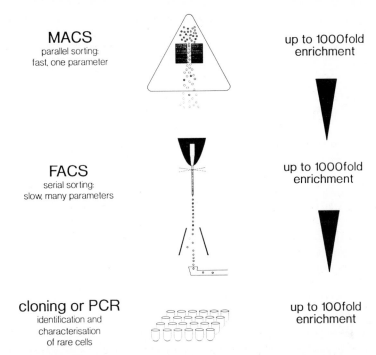

MACS
parallel sorting:
fast, one parameter

up to 1000fold
enrichment

FACS
serial sorting:
slow, many parameters

up to 1000fold
enrichment

cloning or PCR
identification and
characterisation
of rare cells

up to 100fold
enrichment

Figure 2. Efficient selection of rare cells by combining cell sorting methods. This general scheme illustrates how enrichment rates of up to 10^8 fold can be achieved for immunofluorescently labeled rare cells by combination of various cell sorting methods. First, an efficient and fast enrichment of stained cells is obtained by *MACS* which can handle very large cell numbers, as required initially, but is restricted to one parameter. Very good enrichment is obtained with *MACS* because the labeled cells can be sorted out (positive sorting) and directly processed further in a fluorescence-activated cell sorter. The FACS sort is based on serial analysis of several parameters per cell and thus offers very good discrimination but is time consuming which is no longer a problem since the cell number has been reduced in the *MACS* sort. Since FACS can deflect single cells, individual cells can be cloned for further studies or analyzed directly, for example, by polymerase chain reaction, spot enzyme-linked immunosorbent assay, or other single cell based assays.

lized by filtration, or staining reagents can be centrifuged to remove aggregates and microorganisms (5 min in Eppendorf centrifuge)
– Standard cell culture plasticware, such as Eppendorf tubes, sample tubes for sorter, pipettes and tips, syringes, needles
– Nylon meshes with 30–60 μm pore size, cut to pieces of about 2 cm², autoclaved for sterility
– Tissue-culture safe detergent in water
– Ethanol
– Shandon cytocentrifuge or the like
– Fluorescence microscope, slides, coverslips, Fluoromount-G (SBA)
– *MACS* equipment: *MACS* separator and small columns (A1/A2; see Chap. 15)

- FACS equipment: fluorescence-activated cell sorter with sterile sheath fluid (in-line filter), sample filter, and cooling option (see Chap. 16).
- Cell counting device: Neubauer counting chamber or Coulter counter

17.3 Method

Staining the rare cells   Staining must be carefully optimized (see Chap. 3) to maximize discrimination between rare cells and the majority of cells. For *MACS* any background staining can be deleterious.

1. Spin down all cells (usually 10^8–10^{10})
2. Remove supernatant carefully and loosen pellet by flicking.
3. Add titrated staining reagent, for example, biotinylated antibody to the rare cell marker molecule, to a final volume corresponding to the number of rare cells but exceeding the total cell volume at least twofold. Actually, the "negative" majority of cells provides an "exclusion volume" and does not dilute the staining solution. Increasing the staining volume is not critical, but increased concentrations of stain causes background.
4. Stain for 10 min on ice.
5. Dilute the cells with PBS/FCS; spin down (10 min, 300 g)
6. Wash once: remove supernatant, loosen the pellet by flicking, fill up with PBS/FCS, and spin down again.
7. For indirect staining, repeat from step 2 with second reagent, for example, stain the biotinylated first antibody with streptavidin-conjugated superparamagnetic *MACS* microbeads (diluted 1:100 from stock) for 5 min at room temperature; then add titrated amount of phycoerythrin-conjugated streptavidin.

Screening for rare cells: Depending on their frequency, the detection and analysis of rare cells may be difficult by flow cytometry and require extensive live and multiparameter gating to remove unspecifically stained cells, especially dead cells (Fig. 3). An alternative method is detection by fluorescence microscopy, spinning down a defined number of cells on a defined area of a slide, and screening the area for positive cells. This method allows scanning of about 10^6 cells for single positive cells in about 2 h, providing visual information in addition to verify the phenotype.

1. Stain cells as described above
2. Spin 1.2×10^5 cells on each of 10 slides (20% are lost upon centrifugation).
3. Add one drop of Fluoromount-G and coverslip.
4. Analyze by fluorescence microscopy by scanning the slides' areas, counting only positive cells. The number of positive cells on all slides reflects the frequency per 10^6 cells.

Figure 3. Isolation of CD34+ cells from peripheral blood using *MACS* and FACS. Peripheral blood cells were preenriched by Ficoll-Hypaque gradient centrifugation (Chap. 14) and monocytes and NK cells depleted by leucine methyl ester lysis (Chap. 14). The cells were then stained with a biotinylated, titrated CD34 antibody, washed once, stained with streptavidin-conjugated *MACS* beads (diluted 1:100 from stock; Chap. 15) adding 3 µg/ml streptavidin phycoerythrin after 5 min for another 5 minutes, washed and enriched on an A1 *MACS* column (see Chap. 15 for details). The positive cells were eluted from the column outside of the *MACS*, and an aliquot was analyzed and sorted further by flow cytometry (FACStar+) in the presence of propidium iodide. For the *MACS*-sorted cells, large cells and debris were gated out by forward (180°) and side (90°) scatter, dead cells were gated out according to propidium iodide fluorescence in a fluorescence 2 versus fluorescence 3 plot (F2/F3 gating). This gating is crucial for evaluation of the sorting efficiency since unspecifically stained dead cells are enriched together with the rare specific cells. Due to the different fluorescence spectra of phycoerythrin and propidium iodide, the F2/F3 ratio (570/630 nm) is specific for each dye, irrespective of the intensity of staining. Stained cells appear on two parallel diagonals, which are clearly distinct. It should be noted that this approach also al-

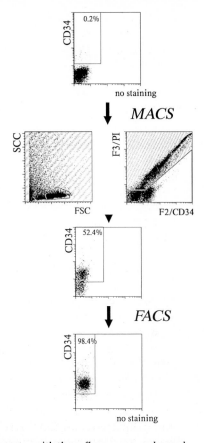

lows use of more than three dyes on a flow cytometer with three fluorescence channels. FACS-sorted cells were reanalyzed on a FACScan using FSC/F3 and F1/SSC to gate on live cells with low side scatter. (The experiment is described in detail in [10].)

The efficiency in terms of recovery of rare cells must be taken into consideration when calculating the number of cells required for sorting. The number of cells prepared for sorting should exceed the number of cells theoretically required at least twofold to compensate for losses during separation, mainly in the FACS.

For *MACS* of rare cells it is important to reduce background staining, to use small columns with just enough capacity for labeled cells, and to use slow flow rates for efficient positive enrichment. **Magnetic enrichment**

1. An A1, A2, or B1 *MACS* column is filled and sterilized from the bottom with ethanol. The ethanol is replaced from the top first with PBS and then PBS/BSA or PBS/FCS, and the column inserted in the *MACS*. The whole procedure can be performed in a sterile workbench.

2. Cells, stained as described above, are filtered through the *MACS* column at low flow rates (0.5–2 ml/min) in two cycles, i.e., cells passing the column are collected and filtered through the column for a second

time. The frequency of cells can be monitored, for example, by checking drops under the microscope from time to time. When the majority of cells have passed the column, the column is washed at higher flow rates, depending on the intensity of staining (high relative fluorescence = high flow rate) until only few cells come with the effluent.

3. The column is then closed at the bottom and carefully removed from the *MACS*. A syringe with 2 ml of PBS/FCS is applied to the side of the stopcock, and the adherent cells are pushed off the wires.

4. The column is inserted back into the *MACS*. The cells are again filtered through the column, first at slow flow rates, then washing at higher rates.

5. When no further cells are washed off the column, the column is eluted outside the MACS by flushing with PBS/FCS from the top. If the elution volume is not too high, the cells can be processed further directly; otherwise it is necessary to spin them down (10 min for 300 *g*). An aliquot is used for analysis of enrichment, either by microscope or by flow cytometry (see Fig. 3).

Fluorescence-activated cell sorting

1. The cell sorter should be set up for optimal performance with respect to optical alignment and calibration. The breakoff point should be stable (resonance frequency) and the deflection streams well focused. Sort recovery and purity should have been checked. Collection tubes should be cooled and a decharging device installed. (All this is described in detail in Chap. 16)

2. The cell path in the sorter is cleaned and sterilized using a tissue-culture compatible detergent (7X, Flow), washing it out again before use with sterile PBS/FCS. The sheath fluid is sterilized by in-line filtration. A sample filter of 30–60 μm pore size should be inserted at the bottom of the sample line as described in Chap. 15.

3. If a substantial number of cells are to be processed, it is advisable to filter the cell suspension beforehand, by aspirating 3 ml cell suspension into a 5 ml-syringe, then fixing a piece of 30- to 60-μm pore size nylon mesh between needle and syringe and slowly passing the cells through the filter into a sample tube. Alternatively, the cell suspension can be filtered directly through nylon gauze into the sorter's sample tube. Large numbers of cells should be sorted in sample medium of similar density as the cells (Chap. 15).

4. Add 3 μl PI stock solution per milliliter of cell suspension, mix, and apply to cell sorter.

5. Analyze negative cells to define sort windows. Gate out dead cells and debris by scatter and PI fluorescence as described in the figure. With *MACS*-enriched cells, i.e., low absolute cell numbers, use low flow rates (under 2000/s), deflect two or three drops per cell and use no sort abort. Processing large cell numbers requires high flow rates (5000–7000) and single-droplet deflection. Whether or not to use sort abort depends on whether the rare cells should be enriched or purified.

6. Check deflection streams, which for rare cells may consist of only occasional droplets instead of a consistent, focused stream, by using a

strong cold light source (see Chap. 15). Switch off the "warm" stream illumination because it may dry out the deflected droplets.

7. Relax and check from time to time the stability of flow rate, breakoff point, and deflection streams. If the number of dead cells and clumps increases, they can be removed by Ficoll-Hypaque (Chap. 14).

8. Sorted cells can be analyzed for purity by flow cytometry. A few thousand deflected cells are sufficient if the PBS/FCS or medium in the collection vessel was filtrated to remove debris particles. Still, gating for cells and against debris is required to evaluate the purity of cells (Fig. 3). Purity and recovery should be protocolled through the sort.

17.4 Tips, Tricks, and Troubleshooting

During long-term sorting from one sample tube, the sedimentation of cells **FACS** may cause severe problems [8, 9]. The flow rate first increases, then **flow rate** decrease, and cells die in the sediment, resulting in substantial overall cell loss. This problem can be solved either by resuspending the cells from time to time or by suspending the cells in medium that is made isopygnic by the addition of Percoll (Pharmacia), to minimize sedimentation [9]. The cells are resuspended in the isopygnic sample medium after the last wash–before filtration through the nylon gauze.

Sterility may become a problem if the cells are sorted for cellular analysis **Sterility** or expansion in tissue culture. Potential ontamination of sorted cells by bacteria can be avoided by adding an antibiotic which is not normally used, such as gentamycin, immediately after sorting. Frequent contaminations of independent sorts may be due to contamination of the sample tubing, the sheath fluid, or common staining reagents. These problems can be eliminated by exchanging the complete tubing system: sample lines, sheath lines, and in-line 0.22-μm sheath filter, as well as replacing the sheath fluid. Autoclavable parts and sample filters should be steam-sterilized. The most frequent source of contamination are staining reagents, especially if they are used by several persons over some time and stored in a refrigerator. Yeast can be eliminated by fungicides such as Nystatin, but it is advisable to separate yeast from cells by Ficoll-Hypaque gradient centrifugation (Chap. 14).

Flushing the sample line with tissue culture compatible detergent (e.g., 0.1% 7X) for a few hours prior to the experiment sterilizes and cleans the sample line, thus preventing adsorbance of cells to the wall of the tubing i.e., bad recovery due to cell loss.

The various aliquots of deflected cells should not be pooled but expanded separately to minimize the risk of contamination.

A general problem with flow-in-air sorters is the introduction of local **Droplet** charge on the plastic of the collection vessel by droplets that do not fall into **repulsion** the collection fluid. This static charge can lead to repulsion of further droplets because it is of the same polarity as the deflected droplets. For rare cells, the deflection streams cannot be focused well, making the problem even worse. To prevent charging, a sterile injection needle is bent, hung onto the inner wall of the collection vessel, and grounded.

Viability No problems with viability have been observed with *MACS*. For enrichment of rare cells, however, one should keep in mind that unspecifically stained dead cells, even if they are rare, are also enriched. Such cells can be removed from the positive fraction by Ficoll-Hypaque gradient centrifugation. For flow analysis and sorting, dead cells must be efficiently gated out, for example by FSC/SSC and F2/F3 gating (Fig. 3).

In FACS the fraction of dead cells rises slowly during a long term sort because cells are dying. This is yet another advantage of *MACS* presorting, which reduces the total sorting time considerably. Nevertheless, in a good buffer more than 80% of most cell types should be alive in the morning if kept on ice overnight. If problems with viability arise, the survival rate should be tested this way, and, if necessary, fresh medium purchased or set up from new stock solutions.

Another problem of FACS is that large and fragile cells may not survive the acceleration at the tip of the nozzle, resulting in bad recovery when comparing numbers of deflected cell and collected cells. A nozzle with larger diameter may improve the situation. Another possible cause of a discrepancy between deflected cell number as indicated by the FACS and the number of collected cells is misalignment of breakoff point or drop delay. This should be aligned and tested carefully before or during the procedure, using standard particles or negative cells. The recovery of deflected cells should be monitored from time to time by counting and/or inspection using the fluorescence microscope. Even a small drift of breakoff point and drop drive frequency can reduce the recovery. Phase setting and small flow impairments may change the direction of the deflection stream or its width. All these parameters should be checked frequently during the sort.

References

(1) - Radbruch, A., Liesegang, B. and Rajewsky, K. (1980) Isolation of variants of mouse myeloma X63 that express changed immunoglobulin class. PNAS USA 77 (5), 2909

(2) - Dangl, J.L., Parks, D.R., Oi, V.T. and Herzenberg, L.A. (1982) Rapid isolation of immunoglobulin isotype switch variants using fluorescence activated cell sorting. Cytometry 2 (6), 395

(3) - Holtkamp, B., Cramer, M., Lemke, H. and Rajewsky, K. (1981) Isolation of a cloned cell line expressing variant H-2K^k using fluorescence-activated cell sorting. Nature 289, 66

(4) - Weichel, W., Liesegang, B., Gehrke, K., Goettlinger, C., Holtkamp, B., Radbruch, A., Stackhouse, T.K. and Rajewsky, K. (1985) Inexpensive upgrading of a FACS I and isolation of rare somatic variants by double-fluorescence sorting. Cytometry 6 (2), 116

(5) - Kavathas, P., Sukhatme, V.P., Herzenberg, L.A. and Parnes, J.R. (1984) Isolation of the gene encoding the human T-lymphocyte differentiation antigen Leu-2/T8 by gene transfer and cDNA subtraction. PNAS USA 81 (24), 7688

(6) - Hombach, J., Leclercq, L., Radbruch, A., Rajewsky, K. and Reth, M. (1988) A novel 34 kD protein co-isolated with the IgM molecule in surface-IgM expressing cells. EMBO J 7 (11), 3451

(7) - Göttlinger, C., Meyer, K.L., Weichel, W., Müller, W., Raftery, B. and Radbruch, A. (1986) Cell-cooling in flow cytometry by peltier elements. Cytometry 7 (3), 295

(8) - Göttlinger, C., Lill, F., Miltenyi, S., Mechtold, B. and Radbruch, A. (1989) Improved flow sorting technology: MACS sorting, cell cooling, flow rate stabilization and computer controlled sorting. Anal. Cell. Pathol. 1 (5-6), 319

(9) - Mechtold, B., Miltenyi, S., Irlenbusch, S.,Göttlinger, C. and Radbruch, A. (1990) Sorting of rare cells, Proc. of 3rd European Cytometry Users meeting, Ghent, Becton-Dickinson Press, pp. 74-81

18 Fluorescence-Activated Chromosome Sorting

J.A. FANTES and D.K. GREEN

18.1 Background

The DNA content of each human chromosome can be measured by flow cytometry, and the results of analyzing a large number of isolated metaphase chromosomes can be accumulated to form a flow karyotype. All the human chromosomes, except for 9–12, can be resolved [1], and small but useful quantities of individual chromosomes can be purified by flow sorting [2]. This material has been used to generate chromosome-specific recombinant DNA libraries which have been used by the scientific communitity as a source of markers linked to genes involved in genetic disease. Recently these libraries have also been used as a source of unique sequences for painting individual chromosomes in metaphase spreads by in situ hybridization [3]. Chromosome sorting has also been used for gene mapping, when cloned DNA probes are localized to a specific chromosome or subregion. This application can involve either filter hybridization [4] or a rapid technique based on the polymerase chain reaction (PCR) [5].

It should be stressed that chromosome sorting is a demanding and time-consuming technique, and alternative methods should always be considered. Several sources of DNA libraries are available commercially [6], and in addition fluorochrome-conjugated libraries for painting are starting to come on the market.

The aim of this section of the course was to give students the opportunity to prepare isolated human metaphase chromosomes from cultured cells and to use the Institute of Genetics flow cytometer to produce a flow karyotype and sort specific chromosomes. These sorted chromosomes were then used in a PCR amplification reaction to localise a known gene to a specific chromosome.

18.2 Material

- 2% detergent solution such as 7X (Flow Laboratories) in filtered distilled water. Warm to 37°C before use.
- Phosphate-buffered saline (PBS) sheath fluid: Oxoid Dulbecco A tablets dissolved in distilled water as instructions. Filter through 0.2 μm filter before use. Store at 4°C and use within 7 days.

- Alternative sheath fluid for PCR: 10 mM Tris, 0.1 M NaCl, 1 mM ethylenediaminetetraacetate (EDTA), pH 7.5. Filter and autoclave before use. Store at 4^{0}C and use within 7 days.
- Complete culture medium: RPMI 1640, 10% fetal calf serum, 12.5 mM 3-[N- Morpholino]propanesulfonic acid (MOPS), 100 U/ml penicillin, 100 µg/ml streptomycin.
- Colcemid: 0.01 mg/ml in distilled water. Filter-sterilize and store at 4°C.
- Polyamine buffer (B1): 15 mM Tris, 0.2 mM spermine, 0.5 mM spermidine, 2 mM EDTA, 0.5 mM ethyleneglycoltetraacetic acid (EGTA), 80 mM KCl, 20 mM NaCl, 14 mM (0.1% v/v) β-mercaptoethanol. Adjust to pH 7.2 with 1 N HCl before adding β-mercaptoethanol. Prepare fresh every week; store at 4°C.
- Polyamine buffer plus digitonin (B2): 0.1% solution of digitonin in B1. Prepare a saturated solution by heating to 37°C and filtering through a 0.2-µm filter to remove any undissolved digitonin. The best source of digitonin is Sigma as some batches of digitonin from other sources are difficult to dissolve.
- 2% Giemsa (improved R66 from Gurr) in buffer pH 6.8 (buffer tablets from Gurr).
- Hoechst 33258 (Calbiochem): 50 µg/ml in distilled water. Store in dark at 4°C.
- Ethidium bromide (Sigma): 1 mg/ml in distilled water. Store in dark at 4°C.
- Chromomycin A3 (Sigma): 1 mg/ml in distilled water. Leave overnight in cold to dissolve. Store in dark at 4^{0}C.
- 0.1 M Magnesium chloride.
- 4,'6'-Diamidino-2-phenyldole (DAPI; Sigma): 50 µg/ml distilled water.
- Taq polymerase and 10 x Taq buffer from Promega. Store at -20°C.
- Deoxyribonucleotides: 50 x stock solution is 10 mM mixture. Store at -20°C.
- Mineral oil from Sigma.

18.3 Methods

The starting material for preparing isolated metaphase chromosomes is a rapidly dividing cell culture with few dead cells and free from bacterial, fungal or mycoplasmal contamination. Cultures with a high mitotic index (>25%) usually give the best preparations, containing mainly single chromosomes with few chromosome aggregates and little debris. The frequency of mitotic cells in a population can be increased by blocking cell division at metaphase with Colcemid, a spindle inhibitor, at the appropriate time in the growth cycle.

The first step in the preparation is to swell the cells in a hypotonic solution to separate the chromosomes and complete spindle breakdown. The cells are then transferred to a chromosome isolation buffer which stabilizes the chromosomes. Finally, the cell wall is broken and chromosomes released into isolation buffer by a combination of detergent action and mechanical force.

Many different chromosome isolation buffers have been described. Use the polyamine buffer developed by Sillar and Young [7], as it is one of the more robust methods and has been used with a wide variety of cells. The buffer contains cationic polyamines such as spermine and spermidine, which stabilize the chromosomes, and chelators such as EDTA and EGTA to inhibit nuclease activity. Highly contracted metaphase chromosomes are obtained which give a good, stable flow karyotype with well-defined peaks. Chromosomes prepared with this buffer contain high molecular weight DNA suitable for most cloning techniques. One disadvantage is that the highly contracted chromosomes are difficult to identify by conventional banding techniques although in situ hybridization with chromosome-specific probes now provides an alternative technique.

Before flow cytometry the chromosomes must be stained with fluorescent dyes; we use a combination of Hoechst 33258, which preferentially binds to AT-rich DNA, with chromomycin A3, which binds to GC-rich DNA. Excitation of H33258 with a UV laser and chromomycin A3 with a 458-nm laser results in a bivariate flow karyotype.

Cell culture   Human chromosomes are usually prepared from lymphoblastoid cell lines (EBV transformed B lymphocytes) or from phytohemagglutinin-stimulated peripheral blood lymphocytes.

1. Set up a culture of lymphoblastoid cells at 3×10^5 viable cells/ml in RPMI 1640 + 10% fetal calf serum.
2. After 30 h add Colcemid to a final concentration of 0.1 µg/ml to block the cells at metaphase.
3. After 16–18 h disperse any cell clumps by gentle pipetting.

Chromosome preparation

1. Take an aliquot of cell suspension for a cell count. Centrifuge cells at 180 g for 10 min. Pour off supernatant and resuspend cells in fresh ice-cold complete medium. Centrifuge cells at 180 g for 10 min; this washing step removes some dead cells and debris.
2. Pour off supernatant and resuspend cells in hypotonic 0.075 M KCl solution to swell the cells. Use 10 ml hypotonic for every 10^7 cells.
3. Incubate cells for 15 min at 37 °C.
4. Remove 250 µl for mitotic index determination: add 5 ml 3:1 methanol:acetic acid and allow to stand for 10 min. Centrifuge at 300 g for 5 min, pour off supernatant, and resuspend pellet in a small volume of fixative. Drop onto clean slide and dry quickly in air. Stain in 2% Giemsa in buffer pH 6.8; wash in distilled water and air dry. Count the number of divisions in 500 cells.
5. After incubation in hypotonic solution (step 3) centrifuge cell suspension at 180 g for 5 min. All further steps should be carried out at 4 °C.
6. Pour off supernatant and resuspend pellet in cold polyamine buffer, (B1), 1 ml/10^7 cells. Centrifuge at 180 g for 5 min.

7. Pour off supernatant. Resuspend pellet in cold polyamine buffer + digitonin (B2), 1 ml/ 10^7 cells.

8. Vortex vigorously for 30–60 s to break cell walls. Check for cell lysis after 30 s vortexing. Monitor cell lysis by phase contrast microscopy or by fluorescence microscopy. For fluorescence microscopy place a drop of chromosome suspension onto a slide previously spread with a drop of fluorochrome such as Hoechst 33258 or ethidium bromide. Place coverslip in position, seal with rubber solution, and examine. Most of the chromosomes should be free and in suspension after 60 s vortexing; further vortexing only causes an increase in chromosome degradation and stickiness.

9. Nuclei should be removed from the chromosome suspension before flow analysis and sorting as they can contaminate sorts. Spin down nuclei at 180 g for 10 min, and transfer supernatant carefully to another tube. Add 1 ml B2 buffer to the pellet and resuspend by a 5 s vortex. Centrifuge at 180 g for 5 min, remove supernatant, and add to first supernatant. Check for the presence of nuclei as described above (step 8).

10. This chromosome suspension can be stored for 2 weeks at 4^0C with little loss of resolution when analyzed by flow cytometry.

1. If the chromosome suspension was prepared from cells with a high **Staining** mitotic index (>30%), dilute 1:1 with fresh B2 buffer before staining.

2. For single fluorochrome analysis add Hoechst 33258 to 0.5 µg/ml or ethidium bromide to 50 µg/ml.

3. For dual fluorochrome analysis add chromomycin A3 to 40 µg/ml, $MgCl_2$ to 1 mM and Hoechst 33258 to 0.5 µg /ml from stock solutions. Leave for at least 1 h for the fluorochromes to equilibrate before analyzing or sorting the chromosomes.

Although chromosomes prepared in polyamine buffer after exposure for **Chromosome** 16 h to Colcemid are condensed, it is possible to band and identify them if **identification** they are swollen and elongated slightly by prior exposure to PBS. Not all the chromosomes on a slide are sufficiently decondensed to give adequate banding for identification, but over 50% should have sufficient bands.

1. Sort 60000 chromosomes into a cold Eppendorf tube containing a quantity of buffer B2 so that the final concentration of sheath fluid after sorting is reduced to 50%. On our machine this quantity of chromosomes is sorted in 0.25 ml of PBS sheath fluid so the sorted chromosomes are finally exposed to 1:1 buffer B2:PBS.

2. Fix chromosomes by adding 40% formaldehyde to give a final concentration of 4%. Leave for 10 min on ice.

3. Spin chromosomes onto alcohol-cleaned slides using a Shandon cytocentrifuge at 150 g for 7 min; 60,000 chromosomes can be split between two slides.

4. Allow slides to air dry. Wash briefly in deionized water and air dry.

5. Fix in 3:1 methanol:acetic acid for 5 min and air dry.

6. Stain in 0.5 μg/ml DAPI in distilled water for 10 min, wash in distilled water, and air dry. Mount in 50:50 antifadent (AFT 10, Citifluor); glycerol.

Flow sorting Chromosome sorting is carried out in most laboratories with a commercial flow cytometer which, depending on usage needs varying amounts of attention before sorting begins. Routine use at least usually guarantees regular performance checks. Time spent on checking the cytometer adjustments and experimenting with the effect which each one has on the coefficient of variation gives credence to the cytometer performance when disappointing results raise doubts about the sample/cytometer overall performance. Fluorescent beads tend to produce contamination of the sorted fractions, and hence it is useful to use an actual sample of suspended chromosomes to adjust the cytometer. Use chromosomes harvested every 2 or 3 weeks from a lymphoblastoid cell line as a "stock suspension" for cytometer adjustment. It is assumed that the mechanics of setting sort windows over the display of accumulated data and the adjustment of droplet cluster, phase, and delay are understood. Invaluable reference sources regarding the technical and biological aspects of chromosome analysis and sorting by flow cytometry can be found [8].

Preparation of flow cytometer Flow cytometers are unavoidably contaminated by the sample material of each experiment. It is therefore essential to flush out the flow cytometer, even following another chromosome sorting session, with solutions which clean but not clog. Before introducing the first cleaning solution (2% 7X) check that the flow nozzle size is suitable for chromosomes–70 μm or less for in air cytometers. This is less critical for enclosed flow chambers in which measurement of the CV is unaffected by nozzle size. After 30 min of warm detergent flushing continue with a further 30 min of distilled water flushing and, finally, introduce the sheath fluid intended for chromosome sorting. Organic solvents should be avoided for flushing purposes since, although these fix any contamination, they also cause contamination to be fixed to the flow tubing only to be dislodged later at the most crucial stage of chromosome sorting. During the flushing process inspect the nozzle with the instrument viewing telescope for small obstructions. Large obstructions cause obvious disruption of the flow stream, small ones only become apparent when suspicious droplet formation and poor chromosome resolution are observed, and this may occur only when one is well into the experiment.

Flow Cytometer Adjustment The experimenter's aim should be to achieve two well aligned laser beams, which give rise to two well shaped and intense fluorescence emission signals from the Hoechst 33258 and chromomycin A3 dyes [9]. Gating on forward and orthogonal light scatter signals does not enhance the resolution of fluorescence signals and should therefore be ignored. The master trigger for capturing the fluorescence intensities of each chromosome should be one of the fluorescence signals. Use a freshly made chromosome suspension as a sample for alignment and begin by centering and focusing the ultra violet laser beam by minimising the width and maximising the intensity of the Hoechst fluorescence. Follow the same procedure for the 458 nm blue laser beam taking care to adjust only those components which effect the alignment of the blue laser and the corresponding fluorescence emission. It is a good idea to practice the fine adjustments, which ideally are one for each laser beam, so that good resolution can be easily maintained throughout the experiment.

During the initial stages of an experiment a period of equilibration between sample and sheath occurs which causes fluctuations in fluorochrome intensities. Stabilization of these fluctuations occurs more rapidly if prior to introduction of the stained chromosome suspension a "dummy" sample containing everything except the chromosomes is passed through the sample stream for 10 min. A reasonable scatter plot of the chromosome fluorescence distribution, required perhaps for filing a permanent record, will result from accumulating 10^5 chromosome signals. An example of Hoechst/chromomycin fluorescence for human chromosomes is shown in Figure 1. **Sample Analysis**

The properties of the charged droplet deflection system should be calibrated before an experiment begins. The most sensitive of these parameters is the droplet delay, the time (measured in fractions of droplets) between signal detection and droplet charging. Optimizing this delay is best performed by sorting a few thousand fluorescent beads on to a microscope slide at a series of droplet delays around the estimated optimum followed by a quick search of the series of "puddles" under the **Sorting Mechanics**

Human Chromosomes

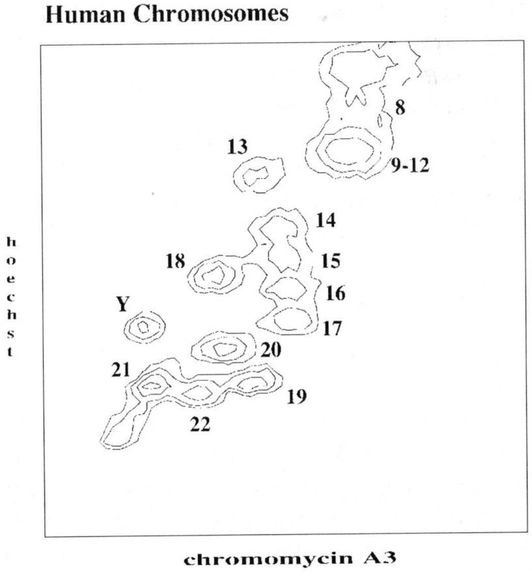

Figure 1. Bivariate flow karyotype of the small human chromosomes (from cell line REN2) stained with Hoechst 33258 and chromomycin A3. Contours have been drawn at 10%, 20%, 40%, and 80% of the maximum frequency registered in the 64 x 64 distribution

fluorescence microscope for the one most dense in sorted beads. From the point of view of catching the sorted chromosomes and the health and safety of the experimenter the undeflected stream is best drawn away by suction into a closed container. The condition for collecting chromosomes into a tube, in terms of which buffer and how much is placed in the tube before sorting, depends on the experiment.

Gene mapping using PCR All the chromosomes in the human karyotype (except 9–12) can be separated and sorted using a flow cytometer, and they provide a useful source of material for mapping genes or DNA fragments. Chromosomes can be sorted directly onto filters and then hybridized with a radioactive labeled DNA probe; only the filter containing chromosomes with a copy of the gene gives a positive signal. Enzymatic amplification of specific DNA sequences with aliquots of flow-sorted chromosomes provides an alternative, faster approach. In this technique two oligonucleotide primers are chosen that hybridise to opposite DNA strands and flank the sequence of interest. The primers hybridize only to chromosomes containing that sequence, and this sequence is then amplified in the PCR, a repetitive series of cycles involving enzymic synthesis of DNA. No amplification will occurs if the chromosomes do not contain that target sequence. The presence or absence of this specific DNA sequence in a fraction of flow-sorted chromosomes can be determined by running the products of the PCR reaction on an agarose gel. Translocation chromosomes, which can be distinguished from the normal chromosomes on a flow karyotype, can be used to localize the DNA sequence to a specific region of a chromosome.

Small quantities of flow-sorted chromosomes can be used for PCR reactions provided they are sorted in a sheath fluid that does not inhibit the PCR reaction; a Tris/NaCl/EDTA buffer is suitable. Five hundred chromosomes of each fraction are sorted directly into tubes that containing buffer for the reaction; the amount of sheath fluid containing 500 chromosomes should be measured for each flow cytometer and should not exceed 2 µl.

Method 1. Sort 500 chromosomes into reaction tubes containing 10 µl of 10x Taq buffer and 70 µl of distilled water.
2. Tap the tube gently to move the droplet containing the chromosomes to the base of the tube. Close the tube and vortex; centrifuge briefly.
3. Freeze the chromosomes in Taq buffer.
4. Add deoxyribonucleotide triphosphates to 0.5 mM and 0.5 M of each primer. Make up to 100 µl with distilled water. Vortex and centrifuge. Cover the surface with 80 µl of mineral oil.
5. Denature at 94°C for 10 min.
6. Add 2.5 units of Taq enzyme. Amplify for 40 cycles using conditions appropriate for the chosen primers.
7. Take 10 µl from each tube and add 1 µl of gel loading buffer.
8. Run on a 2% agarose gel for 3 hours.
9. Stain with ethidium bromide and photograph. Chromosome fractions containing the target sequence will show a bright band of the

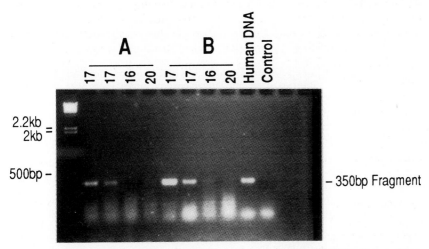

Figure 2. Agarose gel electrophoresis of the products of PCR amplifications using a pair of primers that amplifies a 350-bp fragment of the p53 gene. Fractions of 500 chromosomes (17, 16, and 20) were sorted from chromosome suspensions stabilised either with magnesium (*A*) or polyamines (*B*). The amplified fragment is seen only in fractions containing chromosome 17 and a positive control, human DNA. Left, size markers.

appropriate size; all other chromosome fractions will be negative for this band (Fig. 2).

18.4 Tips, Tricks, and Troubleshooting

1. It is important to treat the cells gently during the preparation; centrifuge at low speeds, resuspend the cell pellet by tapping the tube not by vortexing, and ensure that all the cells are uniformly exposed to hypotonic and buffer solutions by maintaining a single cell suspension.
2. The concentration of cells to buffer B2 at the final stage is important; if the amount of B2 is decreased cell breakage is incomplete.
3. The differential centrifugation step described in step 9 removes most of the contaminating nuclei. Centrifugation of the chromosomes at higher speeds to concentrate them or remove debris increases the number of clumps and degraded chromosomes, giving poor resolution. It is better to start again from a culture with a higher mitotic index.
4. Do not attempt to analyze very concentrated chromosome suspensions. Staining irregularities occur as well as overlapping signals and nozzle blockages. The volumes of buffer B1 and B2 suggested here have been optimized for 1 x 10⁷cells with a mitotic index of 20%–40%. Increase the volume of buffer B2 for preparations with a higher mitotic index.

References

1. Langlois, R.G., Yu, L.-C., Gray, J.W. and Carrano, A.V. (1982) Quantitative karyotyping of human chromosomes by dual beam flow cytometry. Proc. Natl. Acad. Sci. USA 79, 7876-7880.
2. Krumlauf, R., Jeanpierre, M. and Young, B.D. (1982) Construction and characterisation of genomic libraries of specific human chromosomes. Proc. Natl. Acad. Sci. USA 79, 2971-2975.
3. Lichter, P., Cremer, T., Borden, J., Manuelidis, L. and Ward, D.C. (1988) Delineation of individual human chromosomes in metaphase and interphase cells by in situ suppression hybridisation using recombinant DNA libraries. Hum. Genet. 80, 224-234.
4. Lebo, R.V., Gorin, F., Fletterick, R.J., Kao, F., Cheung, M., Bruce, B.D. and Kan, Y. (1984) High resolution chromosome sorting and DNA spot-blot analysis assign McArdle's syndrome to chromosome 11. Science 225, 57-59.
5. Cotter, F., Nasipuri, S., Lam, G. and Young, B.D. (1989) Gene mapping by enzymatic amplification from flow-sorted chromosomes. Genomics 5, 470-474.
6. Van Dilla, M.A. and Deaven, L.L. (1990) Construction of gene libraries for each human chromosome. Cytometry 11, 208-218.
7. Sillar, R. and Young, B.D. (1981) A new method for the preparation of metaphase chromosomes for flow analysis. J. Histochem. Cytochem. 29, 74-78
8. Flow Cytogenetics Ed. Gray, J.W. 1989, Acad. Press.
9. Fantes, J.A. and Green, D.K. (1990) Human chromosome analyses and sorting. In: Methods in Molecular Biology. Vol. 5. (Eds: Pollard, J.W. and Walker, J.M.) Humana Press, 529-542.

19 Flow Cytometry and Sorting of Plant Chromosomes

J. Veuskens, D. Marie, S. Hinnisdaels, and S.C. Brown

19.1 Background

The availability of purified individual chromosomes facilitates the study of molecular properties of eukaryotic genomes. This is mainly based on the reduced DNA amount compared to the total genomic content and on the enrichment of markers or genes located on these chromosomes. Flow cytometry offers the possibility to isolate specific chromosomes and is a routine component of mammalian somatic cell genetics. For instance, one laboratory (Los Alamos National Laboratory) has furnished over 1 000 chromosome-specific libraries.

A flow karyotype is the histogram of relative fluorescence intensities of metaphase chromosomes stained with DNA-specific and/or base pair-specific dyes. Fluorescence intensities of metaphase chromosomes which do not interfere with each other are represented as separated peaks in the flow karyotype only if a sufficient difference in DNA content (generally proportional to relative length) and/or base pair composition exists between the individual chromosomes (Gray and Langlois 1986). Such differences in fluorescense intensities are essential for high purity flow cytometric chromosome sorting (Bartholdi et al. 1987).

The feasibility of flow cytometric sorting of metaphase chromosomes of a particular plant species can be investigated by developing a theoretical model of a flow karyotype based, as a first approximation, on the relative length of the metaphase chromosomes present in the complement (Conia et al. 1989a). Such a model suggests what can reasonably be achieved in terms of purity and yield. Furthermore, a theoretical representation of the total chromosome distribution is an aid for the initial interpretation of the histograms obtained from the flow cytometer.

Flow cytometric sorting of plant chromosomes has so far been reported for *Haplopappus gracilis* cell suspensions (de Laat and Blaas 1984; de Laat and Schel 1986), *Nicotiana plumbaginifolia* (Conia et al. 1989a), *Petunia hybrida* (Conia et al. 1987; 1988) and *Lycopersicon esculentum* cell suspensions (Aramuganathan et al. 1991). A HpaII library has recently been constructed from wheat chromosome 4A enriched preparations obtained upon synchronization of a *Tritium aestivum* (TaKB1) cell culture (personal communication A.R. Leitch and Wang, Norwich, UK). It is improbable that plant flow karyotyping is worth developing as a purely analytical tool, but it could be precious as a preparative step to obtain chromosome specific

DNA. This could then be used for:

– Polymerase chain reaction amplification (Chap. 18)
– Dot-blot hybridization to test DNA probe localization
– Uptake or reinjection into living cells as a method for limited gene transfer through chromosome transplantation
– Restriction, cloning and physical mapping of chromosomal DNA (chromosome specific libraries)
– Study of chromosome specific proteins

Relative to procedures with mammalian cell cultures (Chap. 18) the preparation of plant metaphase chromosomes for both cytological research and flow cytometric sorting is largely hindered by:

– Low frequency of mitotic cells, due to incomplete synchronization and to the lack of simple "tricks" to discard nonmitotic cells
– The presence of a cell wall, such that cells must be reduced to protoplasts with hydrolytic enzymes before releasing metaphase chromosomes
– A tendency to stickiness of the chromosomes after exposure to spindle toxins
– Instability of chromosomes in the current buffers (mammalian chromosome suspensions may be stable for weeks)

To obtain plant metaphase chromosomes different types of biological material are available, such as mesophyll protoplasts, cell suspensions and root meristems. Cultured mesophyll protoplasts usually show a high degree of synchronization in G_0/G_1 phase, leading to a high mitotic index in the first division which can then be chemically blocked. Conia et al. (1987; 1988; 1989a and b), using *Petunia hybrida* protoplasts cultured for 42 h (with 10% metaphase index), thus obtained metaphase chromosome suspensions by simple mechanical disruption in a buffer containing detergent. However, formation of the new cell wall limits the culture period compatible with this isolation step: for instance, division in *Melandrium album* protoplasts is too slow for this strategy. High mitotic indices can also be obtained from cell suspension cultures (the strategy used by de Laat and by Arumuganathan) which are also prone to synchronization by physiological procedures such as nutrient starvation or rapid transfer (Conia et al. 1991). However, chromosome abnormalities frequently appear. Indeed, in the wheat culture used by the Norwich group, no two cells of 40 analyzed contained the same chromosome complement (personal communication A.R. Leitch, Norwich, UK). We believe that genetic and karyotypic instability is a basic argument against using cell suspension cultures even though larger numbers of cells can thus be synchronized and blocked. We have preferred to start from *Agrobacterium rhizogenes* transformed root cultures as a source of dividing cells. The advantages of meristematic root cultures are several: thousands of meristems are available; the root cultures can be maintained on solid medium for years without any effect on chromosome number and chromosome morphology; and the same methods can be rapidly shifted to new varieties.

In this chapter we describe a protocol for the flow cytometric sorting of the Y sex chromosome present in the complement (Ciupercescu et al. 1990; Figs. 1 and 2) of the dioecious plant *M. album* (*Caryophillaceae*). In conjunction, a series of *M. album* plants have been obtained with sexual modifications which are being related to karyological changes in the Y chromosome. Our culture and isolation methodologies should be compatible with this range of material without introducing any further karyological instability. This approach should be seen as an attempt to isolate Y-specific sequences to study at the molecular level factors involved in sex determination and differentiation in higher plants (Ye et al., 1991 and articles in the same number).

19.2 Material

- Petri dishes, 10 cm.
- Gyratory shaker, 50 rpm, 25°C.
- Vacuum pump.
- Stainless steel sieves, 50 μm.
- Conical centrifuge tubes, 15 ml.
- Bench centrifuge.
- Nylon filters, 10 μm mesh. Cut 20 x 20 mm squares of nylon (Monyl, ZBF, Switzerland). Cut across a 1-ml blue pipette tip in two places: 27 mm from its tip (discard this extremity) and 6 mm from the top. Using the latter as a collar, fix the nylon onto the bottom of the central section.
- Culture medium, Gamborg B5.
- Cell cycle synchronization agents (hydroxyurea, aphidicolin, etc). In the present case, 20 mg/ml aphidicolin (Sigma A0781) in dry dimethylsulfoxide (DMSO).
- Antimitotic agents (colchicine, α-bromonaphthalene, dinitroaniline

Synchronization of cells and suspension of chromosomes

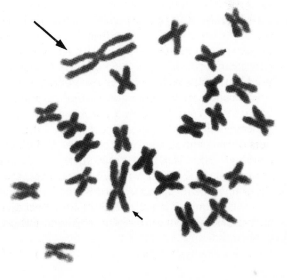

Figure 1. Giemsa-stained metaphase plate of a male line (2n=24, XY) from *Melandrium album*. The X (small arrow) and Y (large arrow) sex chromosomes are 0.14 and 0.21 of the 2C value, the autosomes ranging from 0.06 to 0.09 (Ciupercescu et al., 1990)

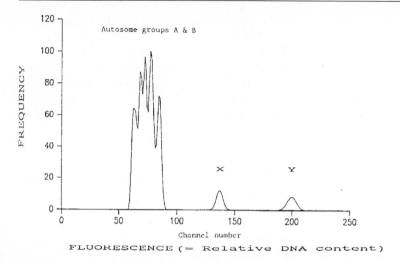

Figure 2. Model flow karyotype for *Melandrium album* (2n=24; male X,Y) where the relative frequency of each chromosome is represented on a histogram by a gaussian distribution at the position according to its relative length in the classical karyotype (Ciupercescu et al. 1990). The sum of all chromosomes is shown, assuming a coefficient of variation (CV) of 2% for the analysis. In reality, the relative lengths do not strictly convert into relative DNA content or into relative fluorescence intensities; minor differences occur even between DAPI and Hoechst 33342 staining, for instance. With basespecific fluorochromes better resolution of the groups A and B (chromosomes 1–11) may be expected in a bi-parametric cytogram [e.g., Hoechst (AT) against mithramycin (GC)]. The larger chromosomes X and Y are conveniently distinct. The CV 2% is realistic. (From Ye et al. 1991)

herbicides such as oryzalin, phosphoric amide herbicides such as amiprophos-methyl, etc). In the present case, 20 mg/ml oryzalin (Elanco, Eli Lilly) in dry DMSO.

– Enzyme mixture [0.15 M sorbitol, 0.05 M Na$_3$ citrate, 2% cellulase Onozuka R10 (Serva), 0.5% – 1% Driselase (Fluka), 0.3% pectolyase Y23 (Yakult), pH 4.8 – 5.0].

– W5 salt solution (154 mM NaCl, 125 mM CaCl$_2$.2H$_2$O, 5 mM KCl, 5 mM glucose, pH 5.7).

– 10 mM hydroxyethylpiperazine ethanesulfonic acid (HEPES) pH 8.0.

– Chromosome suspension buffer A (CSBA: 250 mM KCl, 50 mM MgSO$_4$, 25 mM HEPES pH 8.0, 1.25% Triton X-100, 75 mM β-mer-captoethanol).

– Chromosome suspension buffer B [CSBB: 10 mM HEPES pH 8, 10 mM NaCl, 10 mM KCl, 10 mM spermine, 2.5 mM ethylenediaminetetra-acetate (EDTA), 2.5 mM dithiothreitol (DTT)].

– Nucleic acid specific fluorescent dyes: 1 mg/ml aqueous stocks of bisbenzimide (Hoechst 33342),4'6'-diamidino-2-phenylindole (DAPI), mithramycin in 1 M MgCl$_2$, chromomycin A3 in 1 M MgCl$_2$, ethidium bromide, etc. (Aldrich).

– Formaldehyde, 37% v/v.

- Flow cytometer with stable UV laser (the HeCd laser appears to be inadequate). We use an EPICS V (Coulter) machine with a Spectra- Physics 2025-05 argon laser at 100 mW UV, or 400 mW UV for higher resolution. Emission is collected between 408 and 500 nm. Nozzle size 76 µm. A two-laser cytometer is necessary if AT/GC analysis with Hoechst and mithramycin is to be performed.
- Sheath fluid (75 mM KCl, 2 mM Na$_2$ EDTA, 2mM HEPES pH 8) filtered in-line; 2 l per day.
- Microscopic slides.
- Coverslips, 20 x 20 mm.
- Black filters, Millipore HABP 01300 0.45 µm, as a nonfluorescent antireflex support for epifluorescence observation of sorted material.
- Epifluorescence microscope with appropriate filter sets.
- Film: 400 ASA TMAX black and white, 400 ASA Ektachrome for slides, and 1600 ASA Fujicolor for color prints.
- Copies of classical and model flow karyotypes.
- If possible, chromosome specific probes for dot hybridization to test purity as inclusion or exclusion of specific chromosomes. rDNA probes give a start.

19.3 Method

1. Synchronize young growing roots in early S-phase by adding aphidicolin (final concentration 50 µM) to the culture medium for 24 h. Maintain on shaker.
2. Remove aphidicolin by three subsequent washes of 20 min each with fresh culture medium.
3. Add oryzalin (final concentration 30 µM) to the root cultures and place on a gyratory shaker (50 rpm, 25 °C). Optimal concentration and time of both S and M blocking agents must be determined in advance for each plant species and culture condition; a flow cytometer is useful during this development stage for cell cycle analysis.
4. Collect roots. Using circular culture vessels, we run a scalpel around to isolate the root tips at the perimeter. Transfer root tips to 10 ml enzyme mix. Enhance penetration of enzymes by slight vacuum infiltration for 3–5 min.
5. Incubate on a gyratory shaker (50 rpm, 25 °C). Protoplasts can be harvested when most of the root tip are detached from the rest of the roots (maximum 150 min).
6. Pour gently the protoplast suspension through a stainless steel sieve of 50 µm into a beaker. Rinse the remaining roots with 5 ml 25% W5 solution to collect most of the protoplasts.
7. Transfer the filtered protoplast suspension into 15 ml conical centrifuge tubes and spin at 700 rpm (95 g) for 10 min.

8. Remove supernatant and resuspend pellet in 10 ml 25% W5 solution. Spin as above.

9. Remove supernatant and adjust density of the protoplast suspension to 3 x 10^5 protoplasts/ml.

10. Add 0.5 volumes of 10 mM HEPES pH 8.0 and leave to swell for 10 min.

11. Add 0.2 volumes of prechilled CSBA, mix, and place immediately on ice for 2 min.

12. Add 1 volume of prechilled CSBB, mix, and keep on ice for 2 min.

13. Add DAPI (final concentration 3 µg/ml).

14. Add formaldehyde (final concentration 0.5%), if compatible with subsequent use of DNA.

15. Pass chromosome suspension through 10 µm mesh nylon filter and leave on ice for 15 min before analyzing on flow cytometer.

Flow cytometry and sorting

1. Post the classical and model karyotypes on the wall near the machine! Prepare a model linear karyotype and log karyotype (with 2C and 4C positions included) on transparent film to the same scale as the computer monitor.

2. Clean the cytometer thoroughly; warm the laser 30 min and ensure constant TEM_{00}; equilibrate the sample line with chromosome buffers and dye over 15 min.

3. Parameters: forward-angle light scatter (FALS), linear scale; integral fluorescence intensity on a linear scale (IF) and logarithmic scale (LogIF; 3 decades); peak (or height) fluorescence intensity (PF) on a linear scale.

4. Work profile:

   Cytogram 1: FALS-LogIF used to define "bitmap 1" to include the chromosome region and at least some nuclei as an internal reference of intensities. This bitmap excludes from other histograms a mass of small debris near the origin (starch grains, membrane fragments: low FALS, low LogIF) and large fluorescent material (cell wall fragments, unlysed protoplasts: high FALS, significant LogIF).

   Cytogram 2: IF - PF (gated on FALS and IF) used to define "bitmap 2" to include all objects lying on a close correlation (slope about 45°) between the two parameters. This bitmap excludes from other histograms objects whose fluorescence peak/integral is not as tight as nuclei, micronuclei, or chromosomes. The EPICS V optics are particularly effective in this respect. For instance, swelling or lysis during analysis is immediately evident herein. However, chromosome clusters are not identifiable by this trick.

   Histogram 3: IF (gated on IF, FALS)

   Histogram 4: IF (gated on IF, FALS, map 1, map 2)

   Histogram 5: LogIF (gated on LogIF, FALS)

   Histogram 6: LogIF (gated on IF, FALS, map 1, map 2)

   Histogram 7: PF (gated on IF, FALS, map 1, map 2)

The essential karyotype appears in histograms 4, 6, and 7 (Figs. 3, 4). The sister histograms 3 and 5 enable the operator constantly to monitor the effect and adequacy of the bitmaps, the residual nuclei, and the fine debris. Occasionally, take data in listmode as a check for shift in resolution during analysis.

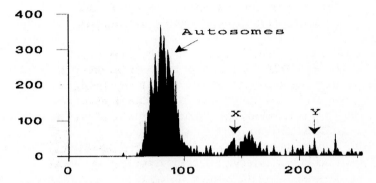

Figure 3. A flow karyotype of male *Melandrium album* (linear scale) established with DAPI (AT specific). Abscissa, fluorescence intensity on a 256 channel scale (related to DNA content); ordinate, frequency. The peak positions and their profile correspond to the predicted ones. The peak areas of the histrogram suggest what sorting confirmed: autosomal clumps lie near or contaminate the X and Y interest zones. Nevertheless, Y can be sorted at 60%–80% purity. Current development is aimed at enhancing metaphase dispersion, increasing chromosome density in suspensions, and notably performing AT versus GC analysis as sex chromosomes will probably then be distinct from autosomal clumps

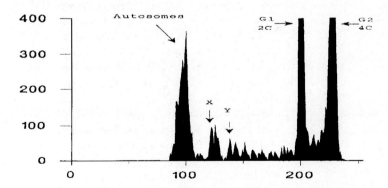

Figure 4. A flow karyotype of male *Melandrium album* (logarithmic scale) established with Hoechst 33342 (AT specific). Abscissa, fluorescence intensity on a 256 channel scale (related to DNA content) after conversion by a logarithmic amplifier (in contrast to the previous figure). This representation is particularly useful when setting up the flow cytometer and when first seeking to identify the relevant peaks. Using the G_1 DNA peak to calibrate the scale, the Y chromosome is then found experimentally at 0.22 times the 2C value and X at 0.15 times (using bisbenzimide Hoechst 33342). In this example, apparently only a small fraction of nuclei were at mitosis

5. If we work at 1000–1500 objects per second, most of which are subcellular particles (debris); a typical analysis may include 10–20 nuclei per second.
6. Find the nuclei in the analysis; set 2C nuclei at channel 200 on the 256-channel LogIF and around channel 150 on the linear IF.
7. Turn up IF and PF by a gain of 5. The Y metaphase chromosome (21% of 2C DNA value) should be near the previous 2C position and the X chromosome at two thirds of the Y position. Lay the "transparent" model karyotypes on the screen; align it with the 2C nuclei on the log scale to see where the various chromosome peaks should (might) be on this scale.
8. Increase FALS to maximize the distinction of putative chromosomes from similar debris, refining the bitmap 1 on cytogram 1 as a wedge (or inverted L) pointing back from the nuclei towards the origin (0,0).
9. Tune the bitmap 2 on cytogram 2. Is the base-line cleaner between the nascent chromosome peaks?
10. Substantially modify the sample flow rate. After allowing equilibration, does this factor affect the resolution of peaks?
11. To understand the various clusters of data, sort 100 events (using anti-coincidence) onto a black Millipore filter, premoistened with CSBB, on a slide; place 5 µl CSBB on the slide to each side of the filter before putting the coverslip for microscopy. (Use filtered medium, clean tweezers slides and coverslips; tissue lint is usually fluorescent.) When sorting larger numbers (e.g., for photography or with nylon for spot hybridization), the filter is placed on a filter holder connecting through a channel in a perspex support to plastic tubing and a syringe so that a very slight depressurization holds the filter in place and evacuates off the substantial liquid that comes with the sorting (Fig. 5). Too much liquid on the filter produces sample dispersion, and too little causes sample instability.
12. These black filters are not compatible with nucleic acid hybridization. For this, sort straight onto routine nylon filters. Sort appropriate controls (e.g., the rest of the chromosome complement or an equivalent number of G_1 nuclei, etc.). Check sorting efficiency by counting the nuclei effectively on a slide after sorting 100.
13. Sort larger numbers into buffer compatible with the intended use of the DNA (e.g., TLES). The tightest sort logic uses FALS, IF (or LogIF), PF, and both bitmaps.

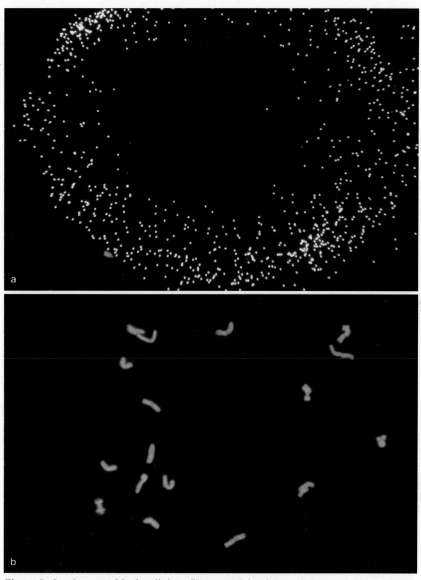

Figure 5. Sorting onto black cellulose filters. **a** *Melandrium album* nuclei from G_0/G_1 phase stained with DAPI. The regularity of the sorting droplet deflection onto a 300 x 400 μm zone has produced this perimeter of concentrated nuclei (x 100). **b** *M. album* Y chromosome. This chromosome is the largest of the complement, with virtually equal arms and represents nearly 21% of the 2C value. Four of the 17 sorted objects are clusters of autosomal chromosomes (x 400)

19.4 Example

The present method (figures) yields 60%–80% purity of Y chromosomes, assessed by morphological criteria after sorting onto black filters. (This chromosome is the largest and most metracentric.) The sole contaminants are chromosome clusters. The best sorting rate approaches one Y chromosome per second.

19.5 Tips, Tricks, and Troubleshooting

Few metaphase chromosomes
- Inappropriate timing and concentration of S and M blocking agents. Optimize synchronization conditions by studying DNA synthesis inhibition and release, and the metaphase block based on flow cytometric cell cycle analysis (Brown et al. 1991) and microscopy. Standardize the state and conditions of growth of plant material. Be aware of formation of micronuclei (de Laat et al. 1987; Verhoeven et al. 1990). With an epifluorescence microscope, check each step of the protoplast preparation from meristematic roots and their lysis.
- Improper hypotonic shock. Allow meristematic root protoplasts to swell without bursting. Check the osmolarity of all the solutions and adjust whenever necessary to obtain maximal distension of the plasma membrane.
- Develop enrichment procedures (absent from this protocol).

Chromosome instability
- Our chromosome suspensions are currently stable for 0.5 – 24 h.
- Use sterile materials and solutions. Contamination induces instability.
- Use formaldehyde, at least while developing an understanding of the flow karyotype.
- Osmotic imbalance.
- Growth imbalance. Browning occurs during the synchronization and blocking steps.

Analysis instability
- Check all cytometry factors (stability of flow rate, laser intensity, temperature).
- Rinse the sample line, then preequilibrate tubing with an identical mixture of buffers and dye (without chromosome material).
- Reduce the size dispersion of objects in the suspension; for example, refilter through 10-μm nylon mesh or reduce nuclei by centrifugation.

Stickiness of metaphase chromosomes by reassociation of their telomeric regions
- Try different mitotic agents which may influence the spindle toxicity.
- Avoid bursting protoplasts during hypotonic swelling when the stabilizing buffer is still inadequate.
- With *M. album*, this has been a major problem. Specific chromosome clusters may be sorted, and the Y-chromosome region is contaminated by doublets and triplets of the much smaller autosomes. A technically

more difficult two-color analysis (see Chap. 18; and Arumuganathan et al. 1991) might circumvent this problem if the Y chromosome has a unique AT/GC ratio. Our protocol yields chromosomes which stain well with the combination mithramycin/Hoechst 33342.

– A second flow karyotype reappears at half the fluorescence intensity values of the primary karyotype. This occasional problem, common during the studies upon *P. hybrida* and *N. plumbaginifolia* (affecting up to one third of all chromosomes, blocked with colchicine) has receded, but it still may be observed. **Formation of mono-chromatid chromosomes**
– Try different mitotic agents. Colchicine seems to favour kinetochore splitting. Although high levels (millimolar) of colchicine are used with plant cells, given the poor affinity of colchicine for plant tubulin *in vitro* compared to mammalian tubulin, some cells often escape through metaphase.

References

– Arumuganathan, K., Slattery, J.P., Tanksley, S.D., Earle, E.D. (1991). Preparation and flow cytometric analysis of metaphase chromosomes of tomato. Theor. Appl. Genet. 82: 101-111.
– Bartholdi, M., Meyne, J., Albright, K., Ledemann, M., Campbell, E., Chritton, D., Deaven, L.L., Cram, L.S. (1987). Chromosome sorting by flow cytometry. Methods in Enzymology 151: 252-267.
– Brown, S., Bergounioux, C., Tallet, S., Marie, D. (1991). Flow cytometry of nuclei for ploidy and cell cycle analysis. In: A laboratory guide for cellular and molecular plant biology. Chapter 5.3 pp. 326-345 (Birkhäuser Verlag, Basel) Eds. Negrutiu, I., Gharti-Chherti, G.B.
– Ciupercescu, D., Veuskens, J., Mouras, A., Ye, D., Briquet, M., Negrutiu, I. (1990). Karyotyping *Melandrium album*, a dioecious plant with heteromorphic sex chromosomes. Genome 33: 556-562.
– Conia, J., Bergounioux, C., Perennes, C., Muller, P., Brown, S., Ganal, P. (1987). Flow cytometric analysis and sorting of plant chromosomes from *Petunia hybrida* protoplasts. Cytometry 8: 500-508.
– Conia, J., Bergounioux, C., Brown, S., Perennes, C., Gadal, P. (1988). Caryotype en flux biparamétrique de *Petunia hybrida*. Tri du chromosome numéro I. C.R. Acad. Sci. Paris 307, III: 609-615.
– Conia, J., Muller, P., Brown, S., Bergounioux, C., Gadal, P. (1989a). Monoparametric models of flow cytometric karyotypes with spreadsheet software. Theor. Appl. Genet. 77: 295-303.
– Conia, J., Muller, P. (1989b). Isolation, classification, and flow cytometric sorting of plant chromosomes. chapt 8, pp 143- 164. In: Flow cytometry: Advanced research and clinical applications. Vol 1, ed. Yen, A., CRC Press, Boca Raton, Florida.
– Conia, J., Alexander, R.G., Wilder, M.E., Richards, K.R., Rice, M.E., Jackson, P.J. (1991). Reversible accumulation of plant suspension cell cultures in G1 phase and subsequent synchronous traverse of the cell cycle. Plant Physiol. 94: 1568-1574.
– de Laat, A.M.M., Blaas, J. (1984). Flow cytometric characterization and sorting of plant chomosomes. Theor. Appl. Genet. 67: 463-467.
– de Laat, A.M.M., Schel, J. (1986). The integrity of metaphase chromosomes of *Haplopappus gracilis* (Nutt.) Gray isolated by flow cytometry. Plant Science 47: 145-151.

– de Laat, A.M.M., Verhoeven, H.A., Sree Ramulu, K., Dijkhuis, P. (1987). Efficient induction by amiprophos-methyl and flow cytometric sorting of micronuclei in *Nicotiana plumbaginifolia*. Planta 172: 473-478.
– Gray, J.W., Langlois, R.G. (1986). Chromosome classification and purification using flow cytometry and sorting. Ann. Rev. Biophys. Bioeng. 15: 195-235.
– Verhoeven, H.A., Ramulu, K.S., Dijkhuis, P. (1990). A comparison of the effect of various spindle toxins on metaphase arrest and formation of micronuclei in cell-suspension cultures of *Nicotiana plumbaginifolia*. Planta 182: 408-414.
– Ye, D., Oliveira, M., Veuskens, J., Wu, Y., Installe, P., Hinnisdaels, S., Truong, A.T., Brown, S., Mouras, A., Negrutiu, I., (1991). Sex determination in the dioecious *Melandrium*. The X/Y chromosome systems allows complementary cloning strategies. Plant Science 80: 93-106.

20 Large Particle Sorting

D.W. GALBRAITH

20.1 Background

The design of jet-in-air flow sorters has centered around the optimal analysis of mammalian cells, particularly those of the lymphoid system, for sensible reasons of commerce. Lymphoid cells have small diameters (ca. 10 µm [2,3]) and are relatively robust. As a general rule, design parameters that are optimal for lymphoid cells are not suited for the analysis and sorting of larger cells (we operationally define large particles as cells or other objects that are larger, often much larger, than 10 µm). Large cells are found in individuals representing all kingdoms of living organisms, including mammalian species. Table 1 lists some examples restricted to the plant and animal kingdoms, although there are many other examples of large cells in the fungi, algae, and protozoans. Since flow cytometry and sorting provide unique ways of examining and quantitatively analyzing cells, it seems inevitable that future interest will grow in the application of these techniques to large particles.

Table 1. Selected examples of some large particles and their sizes

Particle Type	Size (µm)	Reference
Plants		
Broussonetia papyrifera pollen	13.8	[6]
Ambrosia elateior pollen	20.9	[6]
Lycopodium spores	28.6	[6]
Carya illinoensis pollen	51.1	[6]
Zea mays pollen	95.3	[6]
Nicotiana tabacum leaf protoplasts	33.7	[6]
N. tabacum leaf protoplasts	42.3	[6]
N. tabacum cell culture protoplasts	41.2	[6]
Animals		
Adipocyte	94	[3]
Hepatocyte	19	[3]
Megakaryocyte	60	[1]
Multicellular spheroids	41–96	[5]
Purkinje's cells (various species)	16–32	[3]
Oocyte (various species)	60–220 000	[3]
Spermatocyte (various species)	40–215	[3]
Splenic macrophage	25–40	[3]
Splenic wall and reticulum cells	50–120	[3]

The mechanical configuration of the flow cell and the droplet production and deflection processes are illustrated in Figure 1 for a typical jet-in-air sorter (the Coulter EPICS 753). The main elements are a flow cell containing a jeweled orifice out of which the cylindrical fluid jet emerges, an electromechanical (piezoelectric) transducer, pressurized sample introduction and sheath input ports, a droplet charging mechanism, and a droplet observation and deflection area equipped with twin plates maintained at high voltage.

The sample is introduced under pressure into the coaxial laminar stream formed by the sheath fluid. Hydrodynamic focusing centers the sample within the fluid stream both prior to and during its emergence from the flow

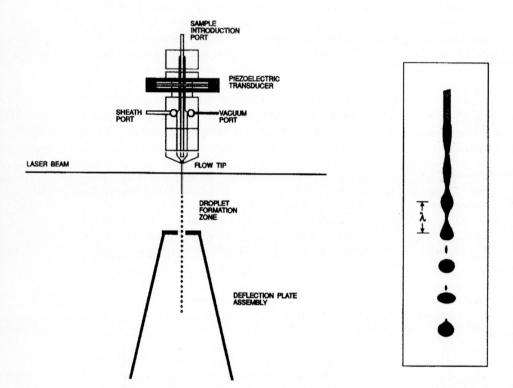

Figure 1. Diagrammatic representation of the EPICS flow cell and droplet deflection assemblies, drawn approximately to scale. Droplet charging involves application of a DC voltage to the stainless-steel sample introduction port. In the standard configuration, the maximum voltage applied to the deflection plate assembly is ± 2 000 V. Insert, illustration of the process of droplet formation at the microscopic level as observed under stroboscopic illumination. The amplification of an undulation possessing a wavelength (λ) longer than the jet circumference results in the precise production of droplets spaced by increments of one wavelength. Satellite droplets are created as a result of separate resealing of the ligament connecting the forming droplet to the fluid jet. In this example, breakage of the ligament has occurred first on the orifice side, resulting in the production of a "fast" satellite, which is accelerated downwards by surface tension and eventually merges with the major droplet ahead of it (redrawn from [9])

cell tip as a high–velocity jet. For most jet-in-air systems, laser interception and signal detection, and droplet formation and sorting occur at sequential points below the flow cell tip (figure). Droplet generation is initiated though actuation of the electromechanical transducer attached to the flow cell and is observed microscopically under stroboscopic illumination. Detection of desired cells or particles activates a timing circuit which places a droplet charging signal on the sample fluid at the instant of formation of the droplet containing the desired cell. This leaves a residual surface charge on the droplet, which then is deflected electrostatically by passage through a fixed electric field. For the Coulter Elite, detection occurs within a quartz flow cell. Droplet formation is achieved after emergence of the fluid from an orifice at the base of the flow cell. Improved collection optics allow use of air-cooled lasers of lower power without compromising sensitivity.

For jet-in-air sorting, droplet formation is a function of the instability of liquid cylinders due to surface tension, and serves to reduce the surface energy of the system. This can be described mathematically as a linear process [4,10] which predicts that a periodic harmonic disturbance imposed on the liquid cylinder will propagate with exponential increase in amplitude only for wavelengths that are greater than the circumference of the cylinder. This process is illustrated in the insert to Figure 1. With liquid cylinders of diameter D_j and an imposed disturbance of wavelength λ, we write:

$$\lambda \times \pi \times D_j \tag{1}$$

For wavelengths shorter than the cylindrical circumference, surface tension is predicted to inhibit waveform amplification and the synchronous production of droplets is not observed. The value of λ_{min} (the minimal wavelength for which amplification will occur) therefore is equal to that of $\pi \times D_j$. Intuitively, the rate of growth of the harmonic disturbance is independent of the velocity with which the liquid cylinder emerges from the flow cell tip. However, the laws of momentum and energy conservation predict that jets emerging from outlet orifices will be smaller than the diameter of the orifice. Under normal operating conditions, jet contraction appears to be significant only for high-speed sorters [7,9].

In flow sorters, the harmonic disturbance is produced through coupling of the electromechanical transducer to the flow cell or to the fluid stream (Fig. 1). The energy of excitation is required only to initiate the process of droplet formation, which subsequently is entirely driven by surface tension, and thus can be very small. Increasing the initial amplitude of the waveform moves the point of droplet formation closer to the flow tip, since less amplification is required before the fluid jet is severed. Empirically, the process of droplet formation is remarkably precise; this is of course critical to the accuracy of sorting.

The relationship between the frequency of excitation of the piezoelectric crystal (f), the velocity of the flow stream (v), and the wavelength of the resultant disturbance (λ) is given by the wave equation:

$$\lambda \times f = v \tag{2}$$

Thus, for a given cylindrical diameter and flow rate, there is a maximal frequency of excitation (f_{max}) of the piezoelectric transducer above which synchronous droplet formation cannot occur. Linear theory also predicts an optimum wavelength ($\lambda_{opt} = D_j \times 4.508$) for which undulation amplification is maximal, and which for a given undulation amplitude minimizes the distance between the flow tip and the point of droplet formation. There is no theoretical lower limit to the frequency of excitation that can be employed for droplet production. However, eventually the rate of amplification becomes less than that of random noise having a wavenumber close to the optimum, leading to the production of random rather than synchronous droplets [4,7]. For commercial instruments, a practical limitation is generally the requirement that droplet formation occur within a fixed distance, that between the flow tip and the point of entry into the droplet deflection region (Fig. 1).

Accuracy of droplet charging is of course critical for accurate sorting. Droplet charging occurs through imposition of an electric potential onto the fluid stream at the time of production of the droplet containing the desired particle. Droplet charging circuits can be adjusted in a variety of ways. The first of these concerns the numbers of droplets that are sorted for each positive detection event. For most machines, sorting can be done in one-, two-, or three-droplet modes which, as their names suggest, describe the numbers of droplets that are charged for each positive detection event. Statistical uncertainty as to the location of the desired particle makes three-droplet sorting the mode of choice when dilute samples are being analyzed. For more concentrated samples, unwanted particles may be found within the time window corresponding to three-droplets. These can be eliminated by switching to two- or one-droplet sort modes or by activating an anticoincidence circuit which aborts sort charging if a nondesired particle is detected within a fixed time window following detection of the desired particle. The second type of adjustment involves establishing the correct time delay between particle detection and droplet charging. This is done by computing the number of wavelengths separating these two points, either by manual observation, or by semi- or fully-automatic procedures (depending on instrument type), thence determining the corresponding time delay. Instruments now permit delay calculations in amounts other than integral wavelength values. The calculated time delay setting is empirically confirmed by sorting of a small, defined number of particles, followed by counting under the microscope to determine sorting efficiency. Most instruments incorporate a further adjustment, that of "phase", which allows droplet charging to be synchronized with droplet generation to eliminate "fanning" of the sorted stream.

For large particle sorting, the first consideration is that of selecting a flow cell tip of the appropriate size and then establishing conditions that give rise to the stable production of droplets above the point of entry into the electrostatic field responsible for droplet deflection. For commercial instruments such as the Coulter EPICS series, flow tips having diameters of 100–200 µm are readily available. Commercial design parameters that

Table 2. Design parameters affecting large particle analysis and sorting compared with their typical implementation for lymphoid cells (modified from [9])

Parameter	Typical design or implementation value on commercial instruments			
	Immunology		Large cells	
Flow cell tip diameter	50	μm	204	μm
Maximum particle size	12	μm	95	μm
Sheath pressure	12	psi	6	psi
Jet velocity	10	m/sec	7.2	m/sec
λ_{min}	140	μm	641	μm
λ_{opt}	200	μm	920	μm
λ_{used}	250	μm	847–2118	μm
Transducer frequency	40	kHz	3.4–8.5	kHz
Break–off length	2.5	mm	8–11	mm

directly or indirectly affect large particle analysis and sorting are given in Table 2, which compares a typical machine configuration used for lymphoid cell sorting with one optimized for sorting large particles. Some design parameters have obvious interactions with particle size (for example, the flow cell tip diameter). Other effects on design result from interactions of previously selected parameters (for example, the pulse timewidth is a function of the particle size, laser beam width, sheath pressure, and flow tip diameter). Yet others (for example, the maximal sort rate) derive from the physical laws underlying the process of droplet production.

As a general rule, adjusting the flow cytometer to give rise to the shortest distance between the flow cell and the point of droplet breakoff is desirable since the available distance for droplet formation is restricted by the fixed positions of the flow tip and deflection plate assembly. From linear analysis, we would expect that conditions satisfying this criterion should involve actuation of the transducer at a frequency which results in an undulation of wavelength at $\lambda_{max.}$. However, two factors empirically influence identification of this optimal frequency. First, the coupling between the piezo-electric transducer and the fluid stream is affected by mechanical resonance and impedance. This means that the efficiency of transfer of energy onto the fluid stream generally is a function of frequency. Second, complex interactions occur between the jet diameter, sorting efficiency, transducer actuation frequency, and particle diameter, particularly as this diameter approaches that of the jet [7]. These result in an interference in droplet formation by large particles that are not centered within the developing droplet. Increasing the wavelength of the imposed undulation reduces this interference since it increases the distance between the large particles and the walls of the developing droplet. Under optimal conditions, transducer frequencies can be selected that permit highly efficient sorting of large particles that are large as 68% of the size of the flow tip [7]. An important point to note is that small particles (such as standard 10-μm fluorescent microspheres) are not sensitive to the interactions experienced by large particles, and therefore they should not be used for optimizing conditions for large particle sorting. The availability of pollen from a variety of plant species spanning size ranges typical of most large cells of interest (Table 1) facilitates the identification of optimal conditions for large particle sorting since pollen is both indestructible and autofluorescent (its fluorescence can be increased through aniline blue staining).

A further point of note concerns achieving adequate droplet deflection. The magnitude of deflection decreases with increasing droplet size, since acceleration of the droplets by the electrostatic field is inversely proportional to the mass of the individual droplets, thus being inversely proportional to the velocity of the flow stream divided by f. From the discussion above, increasing the undulation wavelength in order to minimize particle interference effects therefore decreases droplet deflection, and this eventually may become limiting. The extent of droplet deflection can be increased by increasing the electrostatic field strength (at typical levels, 5 kV/cm, this is well below the spark breakdown limit of air). However, this approach can lead to excessive fanning of the sorted stream. Deflection can also be increased through increasing the time over which the electrostatic field is effective, either by increasing the length of the deflection plate assembly or by decreasing the stream velocity.

It is also important to consider the rate of sorting required by the application. In many cases, maximizing the rate of sorting may be essential. The sort rate is defined solely by f and for droplet production to occur as f is increased, this requires a concomitant increase in the flow rate (E. 1). However, for fixed λ the rate of amplification of the imposed undulation is constant. Thus as the flow rate is increased, the distance between the flow tip and the position of droplet formation also increases. The extent to which this can be accommodated is eventually limited by the distance separating the flow tip from the droplet deflection area.

For the Elite, a further concern might be raised with respect to the pulse width of the signals resulting from the in-quartz detection system. The flow cell, which is square in cross-section, has dimensions ($250 \times 250\,\mu m$) that are considerably larger than those of the standard flow tip. This means that the fluid stream increases in velocity by a factor of about 14-fold as it emerges from the standard tip, from about 0.77 m/s in the sensing area to 10 m/s in the fluid jet. For the 100-μm tip, the corresponding increase is about eight fold, from 1.25 m/s in the sensing area to 10 m/s in the fluid jet. If we approximate the laser beam as an infinitely thin plane, the pulse-width of light scatter or fluorescence emission from a 50-μm particle is 65 μs (for the 76-μm tip) or 40 μs (for the 100-μm tip). These values are larger by at least an order of magnitude than those encountered during standard jet-in-air analysis of lymphocytes (ca. 1 μs). Obviously, the design of the peak detection electronics must accommodate these lengths of pulses. In terms of sorting, the dead time required for particle detection does not appear to be limiting for routine applications. For adequate three-droplet sorting without significant levels of coincidence abort, about one droplet in ten should contain a cell. Thus even at maximal transducer drive frequencies (about 32 kHz at a flow rate of 10 m/s) this corresponds to 3200 events per second or about 312 µs between events.

For sorting large particles and cells, it is clear that an understanding of the theoretical underpinning of the physics of droplet formation is important. However, much information can be achieved through an empirical approach. This series of protocols illustrates how to set up two of the Coulter flow cytometers (The EPICS 753 series and the Elite) for large particle analysis and sorting, through the flow characterization and sorting of pollen, fixed plant protoplasts, and intact, viable plant protoplasts.

20.2 Material

Normal diploid plants of *Nicotiana tabacum* and of *Solanum tuberosum* **Tissue types** were grown vegetatively as sterile shoot cultures in Magenta boxes in agar-solidified basal Murashige/Skoog medium (Gibco Laboratories, Grand Island, New York, USA) containing 3% sucrose (tobacco) and 2% sucrose, 0.4 mg/l thiamine, and 100 mg/l myo-inositol (potato). Pollen and *Lycopodium* spores were obtained from Polysciences (Warrington, Pennsylvania, USA).

All chemicals were obtained from Sigma Chemical. Company (St. Louis. **Chemicals** Missouri, USA) unless otherwise noted. Driselase, Macerase, and Cellulysin were obtained from Calbiochem (La Jolla, California, USA). Mannitol, 2-[*N*-morpholino] ethane sulfonic acid (MES), $CaCl_2$, and aniline blue were obtained from Fisher (Pittsburgh, Pennsylvania, USA); PKH26-GL was obtained from Zynakis (Malvern, Pennsylvania, USA).

General laboratory equipment required for this work includes sterile **Apparatus** plastic petri dishes and plastic 15 and 50 ml centrifuge tubes (Fisher), Millipore Millex-GS (Type SLGS025OS) disposable filters, cheesecloth filters in plastic conical funnels (sterilized by autoclaving), and sterile Pasteur pipettes. A low speed table top centrifuge (up to 1 000 *g*) and a standard fluorescence microscope equipped with epi-illumination and fluorescein excitation/emission filters are also required. Sterility is maintained during tissue culture by performing open manipulations in laminar air-flow sterile cabinet.

20.3 Method

20.3.1 Preparation of protoplasts and pollen

1. Protoplasts are prepared from leaf tissues excised from plantlets approximately 2 weeks after subculture. The leaves (500 mg/dish) are immersed in 10 ml filter-sterilized osmoticum comprising 0.1% driselase, 0.1% macerase and 0.1% cellulysin dissolved in 0.5 M mannitol, 3 mM MES, and 10 mM CaCl$_2$, pH 5.7, contained in a 85 mm diameter sterile plastic petri dish.

2. The leaves are sliced into small pieces (1x10 mm in size) using a scalpel. They are incubated for 12–15 h at room temperature. During this period, the cell walls dissolve, releasing the intact protoplasts. When digestion is complete, the medium should become green and turbid following gentle swirling.

3. The protoplast digest is recovered using sterile cotton-stuffed Pasteur pipettes. It is filtered through two layers of sterile cheesecloth contained in a plastic funnel into plastic conical centrifuge tubes. The protoplasts are pelleted by centrifugation at 100 g for 5 min. The supernatant is discarded.

4. The protoplasts are resuspended in 5 ml of a solution containing 0.5 M sucrose, 3 mM MES, and 10 mM CaCl$_2$, pH 5.7. The suspension is overlaid with 5 ml of protoplast osmoticum to form a step gradient. Care is taken not to disturb the interface. The gradient is centrifuged at 50 g for 5 min.

5. The interface, containing the viable protoplasts, is collected using a Pasteur pipette. It is diluted with five volumes of osmoticum, and the protoplasts are pelleted by centrifugation at 50 g for 5 min.

6. The protoplasts are resuspended in osmoticum to a final concentration of about 10^5/ml. Total protoplast yield is determined by hemocytometry.

7. The proportion of viable protoplasts is found by resuspension of the protoplasts in osmoticum containing 0.1% v/v solution of fluorescein-diacetate (FDA) (1 mg/ml in acetone). Viable protoplasts accumulate fluorescein, and the proportions of these are determined by fluorescence microscopy.

8. If necessary, protoplasts are fixed by incubation for 60 min at 20°C in 2% (w/v) freshly prepared paraformaldehyde dissolved in osmoticum.

9. Pollen (5 mg) is stained by resuspension in 2 ml of a solution [0.1% (w/v)] of aniline blue in phosphate-buffered saline, pH 9.0.

20.3.2 Flow cytometry

20.3.2.1 EPICS System

The EPICS is operated using a 200 μm (nominal internal diameter) flow tip **Instrument** and a sheath pressure of 6 psi. Samples are analyzed at rates typically less **settings** than 500/s. Filter designations on this instrument are as follows: short- (SP) and longpass (LP) filters have a 50% transmittance at the designated wavelength. Below this value, transmittance increases (SP) to 100% or decreases (LP) to zero. Dichroic mirrors (DC) split incident light at the specified wavelength. Light shorter than this wavelength is reflected; longer wavelength light is transmitted. The performance characteristics of the individual filters are provided with the instrumentation documentation but should be confirmed at regular intervals to avoid performance degradation.

1. Sorting intact mesophyll protoplasts. The flow cytometer is operated at a laser wavelength of 457 nm and a power output of 100 mW. Barrier filters LP 510, LP 515, and LP 610 are used (chlorophyll emission occurs above 620 nm). Integral or log-integral red fluorescence, one-parameter histograms are accumulated, and the sort windows are positioned according to the location of the peaks (see, for example, [6]; Fig. 1).

2. Sorting viable mesophyll protoplasts. Viable protoplasts are stained with FDA. The laser is tuned to 457 nm and 100 mW. Barrier filters LP 510 and LP 515 precede the DC 590 dichroic. Two BG 38 filters (Optical Instrument Laboratory, Houston, Texas) are used to screen the green [short wavelength (590) side] photomultiplier tube (PMT). Integral or log-integral green fluorescence, one-parameter histograms are accumulated, and the sort windows are positioned according to the location of the peak.

3. Sorting pollen. The cytometer is operated at a laser wavelength of 514 nm and 200 mW output. Pulse width time of flight (PW-TOF) signals are analyzed based on peak green fluorescence (50% values from the rise to the fall sides of the pulse), using barrier filters LP 530 and LP 540 to screen the PMT. Uniparametric histograms are accumulated [6]. Multiparametric histograms can also be accumulated using forward or 90° angle and integral or peak fluorescence signals, either on linear or logarithmic scales.

4. Sorting fixed protoplasts. The flow cytometer is operated at a laser wavelength of 457 nm and an output of 100 mW, with barrier filters LP 510, LP 515, and LP 590. Uniparametric histograms of log integral red fluorescence are accumulated, and sort windows positioned accordingly.

5. Separation of mesophyll and epidermal protoplasts. The protoplasts are prepared using 0.1% driselase, 0.1% macerase and 0.1% cellulysin dissolved in an ionic osmoticum comprising 0.25 M KCl, 3 mM MES and 10 mM CaCl$_2$, pH 5.7. (Epidermal protoplasts would otherwise float in the mannitol-based osmotica and would be difficult to recover by

centrifugation). Protoplast purification and washing is as described above, except replacing the mannitol osmoticum with the ionic osmoticum at all stages. Protoplasts are FDA-stained as before. The flow cytometer is operated at 457 nm and 100 mW, with filters LP 510, LP 515, DC 590, using two BG 38 filters on the green side and filter LP 610 on the red side. Some signal compensation may be required to eliminate spectral overlap. Bidimensional frequency distributions are accumulated of PW-TOF (based on peak green fluorescence) versus integral red fluorescence (see [6]; Fig. 6).

Instrument alignment   Refer to [7] for a listing of system pressures and appropriate transducer drive frequencies for different flow tip sizes (we routinely use 8 kHz for the 200-µm flow tip and a sheath pressure of 8 psi). The sheath fluid comprises either mannitol (50.5 g/l) and glucose (68.4 g/l) buffered with 3 mM MES, pH 5.7, or solution W5 [8].

Sort alignment

1. Switch to the sort test mode, and adjust the transducer drive frequency to provide a uniform and stable sorted stream for the lowest possible drive amplitude.
2. Count the number of undulations between the laser intersection point and the last attached droplet for an estimate of the charge delay setting in wavelengths.
3. Set up flow analysis of standard particles, using sizes appropriate for the application (paraformaldehyde-fixed protoplasts or pollen). Obtain a flow histogram and set sort windows.
4. Perform a sort matrix analysis to define the charge delay setting that yields maximal sort efficiency. We normally employ the three-droplet sorting mode and a single-cell deposition device (the Coulter Autoclone) to sort 30 particles/well in a 96-well culture plate. We count the numbers recovered after sorting via an inverted light microscope to give the sort efficiency value, varying the charge delay setting around the initial estimate until a sorting efficiency of 100% is obtained.
5. Initiate sorting of protoplasts using the analysis parameters described above and the sort conditions defined by use of the standard particles.

Sterile Sorting

1. For sterile sorting, we utilize a 0.2 µm in-line filter from Pall (Ultipor Type DFA4001ARP). This is autoclaved along with the sample tube, sample pick-up, and sample introduction line.
2. The sheath tanks and sheath lines are cleaned with dilute detergent and are rinsed with 70% ethanol.
3. The sheath fluid is sterilized by Millipore filtration or by autoclaving.
4. Immediately prior to sorting, the sample lines are cleared of residual ethanol by running the system on sample-backflush using sterile sheath fluid. Sorting is performed into 96-well tissue culture microplates that have been half-filled with growth medium. Growth in culture is dependent on protoplast density. Either sufficient protoplasts must be

sorted (about 1 000 per well), or feeder cells must be included in the wells for further protoplast development to proceed.

20.3.2.2 Elite System

In its standard configuration, the Elite is equipped with a 20 mW argon (488-nm) laser and a 10 mW HeNe (633-nm) laser, four PMTs, and a sort-sense flow cell with a 76-μm jet orifice. The recently available 100-μm orifice sort sense flow cell can be used for sorting particles of up to 51 μm in diameter, as described below. We routinely assign the PMTs to 90°angle light scatter (PMT1), green (PMT2), orange (PMT3), and red (PMT4) fluorescence, as illustrated in Fig. 2. Scattered light accesses PMT1 through a 488 nm dichroic (reflected light) and a 488 nm bandpass filter. Light passing through the 488 nm dichroic is filtered through a 488 nm laser blocking filter and successively encounters dichroics splitting at 550 and 625 nm. PMT2 accepts light reflected by the 550-nm dichroic, after passage through a 525-nm bandpass filter (505–545 nm). PMT3 accepts light reflected by the 625-nm dichroic after passage through a 575-nm bandpass filter (555–595 nm). PMT4 accepts light passing through the 625-nm dichroic and through a 675-nm bandpass filter (670–680 nm).

Instrument settings

1. Analysis of pollen and protoplasts. Laser power output is set to 15 mW. Since pollen is autofluorescent (predominantly in the green), it can be readily detected without staining with aniline blue using PMT2 at a high voltage setting of 600 V and an amplification of 7.5. Uniparametric histograms (linear or log-integral signals) are accumulated. Multi-parametric histograms based on forward and 90° angle light scatter (linear or log signals) can also be accumulated. The compressed nature of the peaks using log signals facilitates identification of the different populations. For analysis of potato leaf protoplasts stained with PKH26, biparametric histograms of orange (PMT3) versus red (PMT4) fluorescence are accumulated.

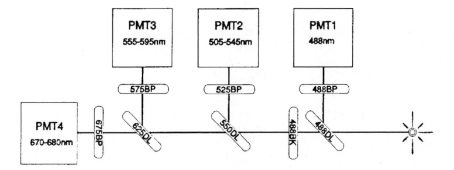

Figure 2. The standard optical arrangement of the filters in the Elite, allowing simultaneous accumulation of three colors of fluorescence as well as forward and 90° angle scatter

2. Sorting pollen and protoplasts. For the 100-μm flow tip, we have employed a system pressure of 8 psi and a drive frequency of 11.5 kHz (for pollen) and 8.9–14 kHz (for protoplasts).

Sort alignment

1. Move the deflection plate assembly as close as possible to the flow cell tip without blocking the incident laser beam (about 5 mm). Examining the flow stream using the videocamera, adjust the transducer drive frequency at constant amplitude to provide as short a droplet breakoff point as possible. Using the sort test mode, adjust the amplitude to give a uniform and stable sorted stream. Prior transducer warm-up is essential (see Sect. 2.6). Lower the deflection plate assembly as required to allow observation of two or three free droplets above the ground plane. The drive amplitude should be less than 20%.

2. Examine the undulations using the videocamera; the last attached droplet should be well rounded on the left side and connected by an obvious ligament to the flow stream on the right side. Adjust the cursor to mark the second well defined undulation to the right of the last attached droplet; mark a second point between the first two free drops above the ground plane. Count the number of droplets between the two cursors and enter this number in the delay calculation program. The optimal delay setting (as determined empirically, described below) should be very close to the calculated delay.

3. Using standard particles approximating the size of the sample to be analyzed, acquire a histogram and set sort windows.

4. Perform a sort matrix analysis to define the delay setting which yields a sort efficiency of 100%. We usually sort 25 particles onto a standard 3 x 1 in. glass slide, and count the particles under a light or fluorescence microscope.

5. After determining the optimum delay, adjust the phase settings and deflection plate assembly high-voltage amplitude to obtain side streams with minimal fanning.

6. Accumulate histograms, establish sort windows, and sort the desired particles.

Sterile sorting

1. Sheath and rinse tanks are cleaned with dilute detergent and filled with 70% ethanol. With sample station on backflush, (place in run mode without tube at sample station), run 70% ethanol from sheath for 15 min.

2. Run shutdown cycle several times with 70% ethanol in rinse tank.

3. Repeat steps 1 and 2 using sterile (autoclaved) water.

4. Replace water with sterile sheath fluid, and repeat steps 1 and 2.

5. Sort into sterile container.

20.4 Examples

20.4.1 Analysis and Sorting of Pollen

Two-dimensional analysis of the 90°angle scatter and autofluorescence signals provided by a mixture of Lycopodium spores with two different types of pollen is given in Figure 3. The paper mulberry pollen and Lycopodium spores comprise single populations. For pecan pollen, two populations are observed. These have similar wide-angle light-scattering properties but differ by approximately six fold in fluorescence (see also [7]). Sorting the low-and high-fluorescence pecan populations (Fig. 4) revealed no obvious difference in subcellular structure as viewed under the

Figure 3. Analysis using the Elite of the 90° scatter (*PMT1*) and green autofluorescence (*PMT2*) signals provided by a mixture of paper mulberry pollen (*A*), Lycopodium spores (*B*), and pecan pollen (*C*). The areas corresponding to the different particles were enclosed by bit-mapped sort regions. The sort efficiencies for all the particles were greater than 98% based on counting of 25 events sorted onto a microscope slide, and the purities were about 95% based on reanalysis of large sorted populations. The contour plots are logarithmically spaced

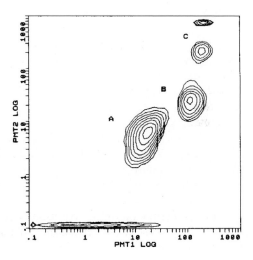

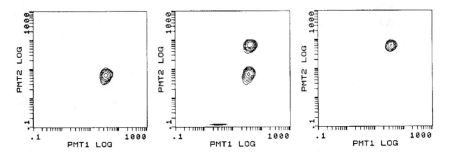

Figure 4. Sorting and reanalysis of high- and low-fluorescence populations of pecan pollen using the Elite. Sorting was based on analysis of log 90° scatter (*PMT1*) and green autofluorescence (*PMT2*) signals (*center panel*), bit-maps placed around the two different populations being separately used for sorting. PMT high voltages were reduced as compared to those used in the previous figure in order to more nearly centralize the two populations of interest within the biparametric plane. Reanalysis was performed on the sorted populations (*left, right panels*) using these same PMT high–voltage and amplification settings

light microscope. The physiological relevance of the low- and high-fluorescence populations remains unclear.

For all the particles, sorting was done with a drop delay of 11.0. The sort efficiencies were greater than 98%. For the 100-μm micron tip, we have calculated a jet velocity of about 9.8 m/s based on measurement of the rate of accumulation of the stream (about 5 ml/min) and assuming that jet contraction is insignificant [7]. For transducer excitation at 11.5 kHz, we predict undulations having a wavelength of 852 μm. The improved efficiency of sorting of the pecan pollen as compared to the situation seen with the 155-μm tip using the EPICS [7] may be related to differences in the orifice geometry.

20.4.2   Analysis and Sorting of Viable Protoplasts

Leaf protoplasts from *S. tuberosum* were prepared and stained with a proprietary lipophilic dye PKH26-GL that has a fluorescence emission maximum in the orange-red region of the spectrum (567 nm). They were analyzed using the Elite (Fig. 5). The initial viability of the population was 90%. After sorting the protoplasts could be clearly distinguished upon reanalysis. However, the viability of the sorted protoplasts dropped to about 50%. This was not affected by the frequency of excitation of the transducer (8.9–14 kHz). It is well known that potato protoplasts are unusually fragile thus this reduction in viability is not unexpected.

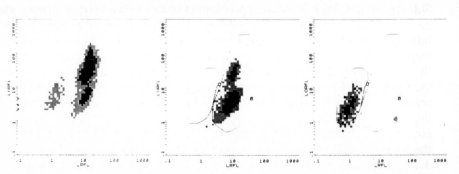

Figure 5. Sorting of viable *S. tuberosum* leaf protoplasts using the Elite equipped with the 100-μm flow tip. Prior to analysis, the protoplasts were stained with PKH26-GL. Biparametric histograms of fluorescence [log orange fluorescence (*LORFL*) versus log red fluorescence (*LRFL*)] were accumulated, revealing at least three protoplast populations (*left panel*). Bitmaps placed around two regions were employed for sorting. Reanalysis of these populations (*center, right panels*) confirmed the precision of the sort process. Most of the contamination in these sorted protoplast populations consisted of subcellular debris

20.5 Modifications

We employ two 2.5 l sheath tanks connected in parallel. This provides **EPICS System** approximately 2–3 h of sorting and analysis between refills with the 200-μm flow tip. Valve VL12, located in the pneumatic system between the sheath pressure safety blow off and the sheath tank, is removed to allow pressure stabilization at lower system pressures. In older EPICS V systems, connecting a switchable ground to pins 2 and 11 on integrated circuit U19 on the sort delay card extends the sort delay to a range of 1–26 wavelengths [7].

No specific modifications to the Elite were required for the efficient sorting **Elite System** of particles as large as 51-μm *Carya* pollen. We have not attempted to sort larger particles. Implementation of sorting with larger flow tips as they become available will probably require increases in the volume of the sheath tank.

20.6 Tips, Tricks, and Troubleshooting

Dependent on the amplitudes and frequencies of transducer activation, the **Satellite droplets** satellite droplets that are observed can be either "fast" or "slow", merging respectively with the droplets ahead of or behind the satellite. Conditions should be established that produce fast satellites only, since these are charged in a manner identical to those droplets with which they will merge.

Side stream fanning is a common problem in setting up for sorting. It may **Side stream fanning** be caused by several different conditions. The transducer drive must be fully warmed up before attempting optimization: set the drive to 70% amplitude and allow to warm up for at least 1 h. Fanning is observed during this period. If the deflection plates are wet, fanning will be severe. Remove the plates, rinse with distilled water, dry completely with methanol, and replace. Excessively high voltage applied to the deflection plate assembly increases the probability of wetting the deflection plates and can also cause fanning due to sort pulse distortion. The high voltage should be adjusted to the lowest percentage consistent with obtaining adequate side stream deflection. Finally, recheck the phase settings to obtain side streams with least fanning.

Check sort efficiencies using smaller particles (e.g., beads, *Broussonetia* **Low sort** pollen) with the same sort settings. If these are sorted with greater **efficiencies of large particles** efficiencies than the large particles, try setting up sorting at longer wavelengths by reducing the transducer drive frequency.

Difficulties in signal acquisition during flow analysis of plant cells Plant cells contain considerable numbers of subcellular organelles, including mitochondria and plastids. When protoplasts are being prepared, a small degree of cellular damage can result in the appearance of a large number of subcellular particles in the protoplast suspension. These particles all scatter light, and some (e.g., chloroplasts) are autofluorescent. This can confuse the neophyte, since the desired particles (the intact protoplasts) can comprise a small proportion of the total particles detected by the flow cytometer. Triggering on fluorescence can be employed to eliminate non-fluorescent, light-scattering particles. Accumulation of log histograms can be helpful in identifying discrete populations. Time-of-flight analysis based on fluorescein diacetate fluorochromasia and accompanied by gating can be used to identify only the viable protoplasts. Finally, use of purification procedures that eliminate nonviable and broken cells limits the contribution of subcellular debris to the particle populations (Due to the impermeable nature and structural strength of the pollen cell wall, subcellular debris is not a problem in the flow analysis and sorting of pollen).

Acknowledgments. I thank Georgina Lambert for valuable technical assistance in the experiments described in the third, fourth, and fifth figures and for help in the preparation of this manuscript. I thank Felix Ayala for providing the Solanum protoplasts employed in the experiments described in the fifth figure. This work was supported by the U.S. National Science Foundation.

References

1. Alberts B, Bray D, Lewis J, Raff M, Roberts K, Watson JD (1989). Molecular biology of the cell, 2nd edition. Garland Press, New York.
2. Altman PL, Katz DD (1976). Cell Biology I. FASEB Press, Bethesda.
3. David H (1977). Quantitative ultrastructural data of animal and human cells. Gustav Fischer Verlag, Stuttgart, New York.
4. Donnelly RJ, Glaberson W (1966). Experiments on the capillary instability of a liquid jet. Proc Roy Soc A 290:547-556.
5. Freyer JP, Wilder ME, Jett JH (1987). Viable sorting of multicellular spheroids by flow cytometry. Cytometry 8:427-436.
6. Galbraith DW, Harkins KR, Jefferson RA (1988). Flow cytometric characterization of the chlorophyll contents and size distributions of plant protoplasts. Cytometry 9:75-83.
7. Harkins KR, Galbraith DW (1987). Factors governing the flow cytometric analysis and sorting of large biological particles. Cytometry 8:60-71.
8. Harkins KR, Jefferson RA, Kavanagh TA, Bevan MW, Galbraith DW (1990). Expression of photosynthesis-related gene fusions is restricted by cell-type in transgenic plants and in transfected protoplasts. Proc Natl Acad Sci USA 87:816-820.
9. Lindmo T, Peters DC, Sweet RG (1990). Flow sorters for biological cells. In: Flow cytometry and cell sorting, 2nd edition (MR Melamed, T Lindmo, ML Mendelsohn, eds.), pp. 145-169, Wiley-Liss, New York.
10. Rayleigh JWS (1876). Notes on hydrodynamics: the contracted vein. Phil Mag 2:441-447.

Part VI Safety

21 Biological and Laser Safety

K. L. Meyer

21.1 Biological Safety

The aim of this short chapter is to remind the reader that, as everywhere else in biomedical research, specimens may be biohazardous. Appropriate measures are necessary to protect the operating personnel and environment. Routine safety precautions are sufficient for "closed" systems such as PAS, PROFILE, FACScan, FACStrak and FACSort.

"Open" flow-in-air systems, such as the sorters of EPIC's and FACStar, produce an aerosol which may contain material from the specimen. Especially for sorting live biohazardous material, special precautions are required.

- Add a sterilizing detergent (Incidin) to the waste reservoir at a concentration sufficient to sterilize the reservoir while it is filled. **Basic precautions for closed systems**
- Flush the tubings with 0.5% sodium hypochloride (Chlorix, fresh dilution) and then with a 0.1% detergent solution such as 7X (Flow) or another tissue-culture compatible detergent. This should be done routinely every day. If debris piles up in the sample tubing, cleaning it out promptly may cause severe clogging.
- Follow the general guidelines for work with biohazarous material. Such guidelines provided, for example, in *Biosafety in Microbiological and Biomedical Laboratories* [1] and *BioTechnologie* [2].

- If possible, fix the cells for example, in 1% paraformaldehyde/phospate-buffered saline overnight. Sorting of live biohazardous material should be carried out only under extreme precautions, placing the sorting chamber in a sterile downflow cabinet licensed for that type of biohazardous material and, in addition, keeping the sort chamber under low pressure. Also, operating personnel should wear gloves and masks. **Additional precautions for "flow in air" systems**
- Low pressure should in any case be applied to the sort chamber to get rid of the aerosol, even for fixed material.
- Decontaminate the surface of the whole sort chamber in addition to tubing and nozzle.

21.2 Laser Safety

Most flow cytometers are run with one or more lasers. All of these are more or less dangerous. In terms of safety, lasers are distinguished into five classes: class 1, class 2, class 3A, class 3B, and class 4 (Fig. 1). "Closed" systems are normally provided with a class 3B laser operating, for example, with 15 mW at 488 nm. (A 1 mW laser beam is equivalent to the power of the bright sun.)

"Flow-in-air" systems are generally equipped with lasers of the highest class, class 4. These lasers are normally used so as to bring up to 150 mW to the illumination point, i.e., still in the range of class 3B lasers, although they may and sometimes are be used at higher power. The direct laser beam of class 3B and class 4 lasers cause permanent damage of the eye. Direct light of class 4 lasers cause damage to the skin in addition [3, 4]. Diffuse radiation from class 3B lasers is not dangerous to the eye at a distance of more than 13 cm for a period of less than 10 seconds [3]. Any exposure to class 4 laser light damages eye and skin. Flow cytometers are normally provided with laser safety covers and interlock shutters to make them safe. Especially for flow-in-air systems, laser safety devices are frequently removed in the course of modifications or because they disturb handling of the instrument. In this case, the operator should know exactly what she/he is doing and should take care that laser safety is maintained. Warning symbols must always be posted in areas of potential danger (Figs. 2 and 3). For a list of safety measurements see Table 1.

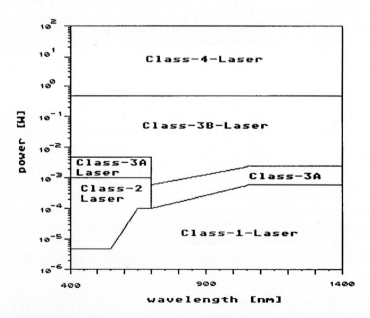

Figures 1. Simplified classification of continuous wave (CW) lasers in terms of the wavelength and power (according to [5,6])

Table 1. Safety measures for lasers of classes 3 B and 4 (from [3,4])

Laser type	Power spec.	Power used	Precautions
Class 3B			– Use dark, nonreflecting tools with small radius, avoid plain surfaces.
			– Separate the laser work area spatially.
			– Appropriate signs, such as laser warning labels, must be posted in front of direct access to the laser radiation.
FACScan	15 mW	15 mW	– Use optical protection filters or glasses.
			– Never look directly into a laser beam.
EPICS			– Never suppress the closing reflex of the eyelid; this protects the eye from scattered laser radiation.
PROFILE	15 mW	15 mW	– Maximum daily exposure to laser radiation should not exceed 8 h.
			– Every person operating the instrument with direct access to laser radiation must be instructed. (a)
Class 4			– All surfaces in the working room should be nonreflecting.
Coherent:			– Avoid plane surfaces on top of the lasers to protect the lasers from any liquid.
Innova 70-2	2W	0.1–2 W	– Separate the laser working room as a special facility.
70-3	3 W	0.1–3 W	– Avoid exposure to inflammable material or gases such as oxygen, ether, gasoline, and alcohol.
70-4	4 W	0.1–4 W	– The laser area must be classified as a controlled area with limited access to authorized persons.
Innova 304	4 W	0.1–4 W	– The laser itself and the laser area (working room) must be locked, with restricted access to authorized persons (operate
305	5 W	0.1–5 W	the laser in a room with the door clearly marked and locked).
306	6 W	0.1–6 W	– Appropriate signs and signals, such as laser warning lamps and warning labels, must be noticeable and posted at the entrance to the laser room.
Spectra-Physiks:			
2020-03	3 W	0.1–3 W	– Do the beam adjustment with low laser power and at a different height from the eye.
2020-05	5 W	0.1–5 W	– Continuity in adjustment and running the instrument must be given.
2025-03 etc.	3 W	0.1–3 W	– Only properly prepared persons should be permitted to operate the instrument, and they must be instructed. (a)
UV lasers	0.2–0.5 W	0.05–0.5 W	– The UV lines (e.g. argon laser) are longer than 320 nm, which means that scattered UV laser radiation can cause eye and skin damage (sunburn), but it is probably not carcinogenic.

(a) according to [3,4] or alike depending on the contry.

Figure 2. Laser symbol. This label must to be posted at the entrance to the laser working room and at any access to laser radiation

Figure 3. Class 4 laser warning label. This or a similar label must be posted near to the direct access of class 4 laser radiation

Note. For general physical safety follow the statements of instruction manual for the flow cytometer.

References

1. Biosafety in microbiological and biomedical laboratories. U.S. Department of Health and Human Services, March 1984, HHS Publication No. (CDC) 84-8395
2. BioTechnologie., polycom Verlag, Braunschweig, FRG, 1987, ISBN-No. 3-9801406-1-X
3. Unfallverhütungsvorschrift VBG 93: Unfallverhütungsvorschrift Laserstrahlung (VBG 93), Carl Heymanns Verlag, Köln, FRG, 1988. Order No. VBG 93 and VBG 93 Durchführungsanweisungen
4. ANSI Standard Z136.1, Safe use of lasers: American National Standard for the Safe Use Of Lasers. American National Standards Institute, 1980
5. DIN VDE 0837: Strahlungssicherheit von Lasereinrichtungen, Klassifizierungen von Anlagen, Anforderungen, Benutzer-Richtlinien; Deutsche Elektronische Kommission im DIN und VDE
6. ICE Publ. 825 (1. ed. 1984): Radiation safety of laser products, equipment classification, requirements and user's guide. CEI Genéve

Glossary

B. MECHTOLD and A. RADBRUCH

Absorption: The taking up of light, sometimes only of certain wavelengths (absorption wavelength) and transforming it either to other forms of energy or to light of other wavelenghts (fluorescence).

Acquisition: The recording of data in flow cytometry.

Alignment microscope: Serves to monitor the nozzle, flow stream, and breakoff point, as illuminated by a stroboscope in drop-drive frequency, to measure distances (=time) between laser interception and breakoff point, and to calculate the drop delay.

Anticoincidence circuit: If cells pass the point of analysis at too short a distance, both cells can give an artificial signal, or the second cell is not measured correctly because the system is not yet ready (analysis abort). For cell sorting, cells following each other too closely to be packed into independent droplets, or differentially sorted droplets, may be included for the sake of recovery, or not, for optimal purity (sort abort).

Allophycocyanin (APC): A fluorescent phycobiliprotein isolated from the cyanobacterium *Anabena variabilis*. The peak of fluorescence emission is 660 nm; the peak of absorption is close to 650 nm. For flow cytometry it can be excited by dye lasers (ca. 560 nm) or helium/neon lasers (632 nm).

Argon laser: The most frequently used illuminating light source in flow cytometry. Useful for UV excitation of Hoechst dyes (DNA of live cells, bromodeoxyuridine/Hoechst quenching) and indo-1 measurements: useful for excitation of most immunofluorescence dyes, propidium iodide (DNA), and fluo-3 (Ca flux) with 488-nm blue light; sometimes also used for pumping dye lasers with 514-nm light.

Autofluorescence: Endogeneous fluorescence, a problem with fibroblasts and fixed cells, can be compensated by measuring it as an independent parameter.

Bandpass filter (BP): An optical filter transmitting a small spectral range of light.

Barrier filter: An optical filter blocking a small spectral range.

Channel: The numberic value of a parameter, i.e. the digitalized signal intensity. The range is 256–1024 channels.

Cluster designation (CD): The nomenclature for leukocyte surface antigens. Listed in : Leucocyte typing IV by W. Knapp, Oxford University Press, 1989 and a poster by Janssen Biochimica, Belgium.

Compensation: The subtraction of cross-talk between photodetectors.

Contour plot: A plot showing two parameters coordinated, with contour lines circumventing areas containing equal densities of events in linear, log, or probability fashion.

Coefficient of variation (CV): The standard deviation (channel number) divided by the mean fluorescence of the fluorescence distribution, x 100.

Dichroic mirror (DC): A mirror which transmits and reflects light according to wavelength.

Dot plot: A plot showing two parameters coordinated, with each event plotted as a dot according to the intensity of the parameters.

Drop delay: Distance (time) required by a cell to fly from the point of analysis to the breakoff point, i.e., from analysis to sorting, in flow-in-air sorters.

Fluorescence: The property of molecules to absorb light of a particular wavelength and emit light of a longer wavelength.

Forward scatter (FSC): Light scattered by a cell in the direction of the excitation light beam as the cell passes the beam.

Gate: Intensity thresholds defining a region of intensity of a particular flow cytometric parameter, used for further analysis or sorting.

HeCd laser: Helium-cadmium ion laser.

Live gate: A gate set for the recording of cells.

Longpass (LP): An optical filter blocking all light shorter than a certain wavelength

Mean: The sum (arithmetic, linear amplification)) or product (geometric, logarithmic amplification) of all signals, divided by the number of signals.

Median: The channel at which 50% of cells lie below and 50% above.

Orthogonal light scatter: Light scattered to the side (see forward scatter).

Peltier element: An electronic element generating a temperature gradient from current. Useful for controlling temperature in flow, e.g., to cool cells.

Photomultiplier tube (PMT): A photodetecto, with photons generating a cascade of electrons to generate a measurable current.

Propidium iodide (PI): A red-fluorescent dye staining DNA; a marker for dead cells because it is excluded quantitatively by live cells.

Piezoelectric: An electronic element converting electronic frequencies into mechanical vibrations of the same frequency. Used to generate a stable breakoff point and a defined number of droplets in flow-in-air systems.

Shortpass (SP): An optical filter that blocks all light of wavelengths longer than a particular wavelength.

Side scatter (SSC): Light scattered to the side as the cell passes through the laser beam.

Threshold trigger: The minimum signal intensity to trigger data processing.

List of Suppliers

B. MECHTOLD

Suppliers

Products

Aldrich-Chemie GmbH & Co. KG
Postfach 1120
Riedstr. 2
W 7924 Steinheim
Tel.:0049. 7329. 87-110 0130. 7272
FAX:0049. 7329. 87-139

Dyes and chemicals

Bachem Biochemica GmbH
Lessingstr. 26
W 6900 Heidelberg 1
Tel.:0049. 6221. 163091
FAX:0049. 6221. 21442

Biochemicals

Bachem Bioscience INC.
3700 Market St.
Philadelphia PA 19104, USA
Tel.:001. 215. 387-0011
FAX:001. 215. 387-1170

s.a.

Balzers GmbH
Fürstentum Liechtenstein
Tel.: 075. 44111

Interference filters

Becton Dickinson GmbH
Tullastr. 8-12
W 6900 Heidelberg 1
Tel.:0049. 6221. 305-0
FAX:0049. 6221. 305-216

Flow cytometers:
(FACSort, FACStar
FACStrack, and FACScan)
and antibodies

**Becton Dickinson
Immunocytometry Systems**
2350 Qume Dr.
San Jose CA 95131-1807, USA
Tel.:001. 408. 432-9475
FAX:001. 408. 954-2009

s.a

Boehringer Mannheim GmbH
Biochemica
Postfach 310120
Sandhofer Str. 116
W 6800 Mannheim 31
Tel.:0049. 621. 759-8545
FAX:0049. 621. 759-8509

Biochemicals and antibodies

Suppliers

Products

Boehringer Mannheim Biochemicals
P. O. Box 50414
9115 Hague Rd.
Indianapolis IN 46250-0414, USA
Tel.:001. 317. 849-9350
FAX:001. 317. 576-4065

s.a.

Calbiochem Novabiochem GmbH
Postfach 1167
Lisztweg 1
W 6232 Bad Soden/TS
Tel.:0049. 6196. 63955
FAX:0049. 6196. 62361

Antibodies, biochemicals, and
Hoechst dyes

Calbiochem Corporation
P. O. Box 12087
San Diego CA 92112-4180, USA
Tel.:001. 619-450-9600
FAX:800-776-0999

s.a.

Coherent, Inc.
Laser Products Div.
3210 Porter Dr.
Palo Alto, CA 94304, USA
Tel.:001. 415. 493-2111

Laser manufactures

Coulter Electronics GmbH
Postfach 547
Gahlingspfad 53
W 4150 Krefeld 1
Tel.:0049. 2151. 818-180
FAX:0049. 2151. 802-000

Flow cytometers:
(EPICS, Profile, and ELITE)
and antibodies

Cyanotech Corporation
P. O. Box 4384
Kailua-Kona HI 96745, USA
Tel.:001. 808. 326-1353
FAX:001. 808. 329-3597

Antibodies and phycobiliproteins

Dow Elanco
Derquinter 92
B 2018 Antwerpen, Belgium
Tel.:03.247-4225

Antimitotic agents

DUNN Labortechnik GmbH
Postfach 1104
Zurheiden 6
W 5464 Asbach
Tel.:0049. 2683. 43306
FAX:0049. 2683. 42776

Dyes and antibodies

Elanco Products Company → **Eli Lilly**
Eli Lilly and Company → **Dow Elanco**

Eppendorf Gerätebau　　　　　　　　Laboratory equipment
Netheler + Hinz GmbH
Postfach 650670
Barkhausenweg 1
W 2000 Hamburg 65
Tel.:0049. 40. 53801-0
FAX:0049. 53801-556

Erbslöh　　　　　　　　　　　　　　Nylon gaze
Kaiserstr. 5
W 4000 Düsseldorf
Tel.:0049. 211. 39001-0
FAX:0049. 211. 39001-21

EXCITON, Inc.　　　　　　　　　　Antibodies
P. O. Box 31126
Overlook Station, Dayton
Ohio 45431, USA
Tel.:001. 513. 252-2989
FAX:001. 513. 258-3937

Falcon → **Becton Dickinson**　　　　Tissue culture, tubes and others

Fisher Scientific Company　　　　　Chemicals
Pittsburgh - Pennsylvania USA
Tel.:001. 412. 963-3300

Flow → **Serva**

Flow Cytometry Standards Corporation　Calibration particles (beads)
P. O. Box 12621
Research Triangle Park NC 27709, USA
Tel.:919. 967-9345
FAX:919. 967-7325

Fluka Feinchemikalien GmbH　　　Chemicals
Postfach 1346
Messerschmittstr. 17
W 7910 Neu-Ulm
Tel.:0049. 31. 70111 0130. 2341
FAX:0049. 731. 75044

Fluka Chemical Corporation　　　　s.a.
980 So. Second St.
Ronkonkoma NY 11779-7238, USA
Tel.:516. 467-0980
FAX:516. 467-0663

Gibco BRL AG　　　　　　　　　Biochemicals
Postfach 1212
Dieselstr. 5
W 7514 Eggenstein-Leopoldshafen 1
Tel.:0049. 721. 7804-0
FAX:0049. 721. 780499

Suppliers

Products

Gibco BRL
Life Technologies INC.
3175 Staley Rd.
Grand Island NY 14072, USA
Tel.:001. 716. 773-0700
FAX:800. 331-2286

s.a.

Gurr
BDH Laboratory Supplies
Broom Road
Poole BH12 4 NN, England
Tel.:-202. 745520
FAX:-202. 745520

Dyes

Heraeus Instruments GmbH
Produktbereich Thermotech
Postfach 1563
Heraeusstr. 12-14
W 6450 Hanau 1
Tel.:0049. 6181. 35-5190
FAX:0049. 6181. 35-784

Laboratory equipment

IBL Research Products INC.
67 Rogers St.
Cambridge MA 02142, USA
Tel.:617. 868-0077
FAX:617. 661-6341

Antibodies

Immunotech (Dianova-Immunotech)
Postfach 101705
Raboisen 5
W 2000 Hamburg 1
Tel.:0049. 40. 323074
FAX:0049. 40. 322190

Antibodies

Immunotech S.A.
Luminy case 915
F 13009 Marseille cedex 9, France
Tel.:0033. 91758200
FAX:0033. 91412794

Antibodies

Janssen Biochimica
J. Pharmaceuticalaan 3
B 2440 Geel, Belgium
Tel.:0032. 14. 604290
FAX:0032. 14. 604220

Biochemicals

Labomatic GmbH
Postfach 469
W 6920 Sinsheim
Tel.:0049. 7261. 64903

Silicon and tygon tubes

Laseroptik GmbH
W 3008 Garbsen 8
Tel.:0049. 5131. 2188

Optical supply:
Special surface of filters

Merck, E.
Postfach 4119
Frankfurter Str. 250
W 6100 Darmstadt 1
Tel.:0049. 6151. 72-0
FAX:0049. 6151. 72-7521

Biochemicals

Millipore GmbH
Hauptstr. 71-79
W 6236 Eschborn
Tel.:0049. 6196. 494-0
FAX:0049. 6196. 43901

Equipment for filtrations
of gases and fluids

Miltenyi Biotec GmbH
Friedrich-Ebert-Str. 68
W 5060 Bergisch Gladbach 1
Tel.:0049. 2204. 8096
FAX:0049. 2204. 85197

Magnetic cell sorters (MACS),
beads, and colums

Molecular Probes INC.
4849 Pitchford Ave.
Eugene OR 97402, USA
Tel.:001. 503. 344-3007
FAX:001. 503. 344-6504

Biochemicals and antibodies

Monyl
Zurich Bolting Cloth Mfg. Co. Ltd.
Ch 8803 Rüschlikon
Switzerland

Nylon gaze

Newport Corporation
18235 Mt. Baldy Cir.
P. O. Box 8020
Fountain Valley, CA 92728-8020, USA
Tel.:001. 714. 963-9811

Optical supply:
optical tables and mounts

Omega Optical INC.
3 Grove St.
P. O. Box 573
Brattleboro, VT 05301, USA
Tel.:001. 802. 254-2690

Interference filters

Oriel GmbH (LOT)
Im Tiefen See 58
W 6100 Darmstadt
Tel.:0049. 6151. 82076(8806-0)
FAX:0049. 6151. 419602 (84173)

Optical supply:
Arc lamp systems and power supplies

Ortho Diagnostic Systems GmbH
c/o Dr. Molter GmbH
Postfach 1320
Karl-Landsteiner Str. 1
W 6903 Neckargemünd
Tel.:0049. 6223. 77-0
FAX:0049. 6223. 77-278

Flow cytometers:
(CYTORON) and antibodies

Suppliers

Products

Ortho-Mune Monoclonal Antibody
Div. Ortho Diagnostic Systems Inc.
Route 202
Raritan NJ 08869, USA
Tel.:001. 201. 218-1300
FAX:001. 201. 218-8582

Dyes, biochemicals and
antibodies

Pall Filtrationstechnik GmbH
Postfach 102120
Philipp-Reis-Str. 6
W 6072 Dreieich 1
Tel.:0049. 6103. 307-0
FAX:0049. 6103. 34037

Equipment for filtration of fluids
(large volume filtration)

Partec GmbH
Hüferstr. 73-79
W 4400 Münster
Tel.:0049. 251. 80078
FAX:0049. 251. 82979

Flow cytometers:
(PAS-machines and
Cell Analyzer CA)

Pharmacia Biosystems GmbH
Postfach 5480
Munzinger Str. 9
W 7800 Freiburg 1
Tel.:0049. 761. 4903-30
FAX:0049. 761.4903-159

Laboratory equipment

Pharmacia LKB Biotechnology AB
751 82 Uppsala, Sweden
Tel.:46. 18163000
FAX:46. 18143820

Biochemicals

Pharmingen
11555 Sorrento Valley Rd. #E
San Diego CA 92121, USA
Tel.:001. 619. 792-5730
FAX:001. 619. 792-5238

Antibodies

Phoenix Flow Systems
11575 Sorrento Valley Road
Suite 208
San Diego, CA 92121, USA
Tel.:001. 619. 453-5095
FAX:001. 619. 259-52668

Software products for the PC:
MultiCycle-DNA analysis software
MultiReport-Clinical flow cytometry
Report generator software
Multi2D-Bivariate data analysis software
MultiTime-Kinetic versus time data
analysis
Software
ListView-list mode analysis software
under Windows 3.1
LabReporter-Clinical report generator
software for Windows 3.1
ACQCYTE-Data acquisition software
for flow cytometry

Pierce Europe B.V.
P. O. Box 1512
3260 BA oud Beijerland, Holland
Tel.:31. 1860. 19277
FAX:31. 1860. 19179

Antibodies

Pierce
P. O. Box 117
Rockford, IL 61105, USA
Tel.:001. 815. 968-0747
FAX:001. 815. 968-8148

s.a.

Polysciences, INC.
400 Valley Road
Warrington, PA 18976-2590, USA
Tel.:   001. 215. 343-6484
          800. 523-2575
FAX: 001. 215. 343-0214

Dyes, calibration particles and
enzyme substrates

Polysciences Limited
Niederlassung St. Goar
Postfach 64
Ulmenhof 28
W 5401 St. Goar
Tel.:0049. 6741. 2081
FAX:0049. 6741. 2089

s.a.

Promega Ltd:
Epsilon House
Enterprise Road
Chilworth Centre
Southampton S01 7NS, England
Tel.:0044. 703. 760225
FAX:0044. 703. 7607014

Enzymes and biochemicals

SCHOTT Glaswerke
Postfach 2480
W 6500 Mainz
Tel.:0049. 6131. 66-0
FAX:0049. 6131. 662003

Optical filters

Schweizerische Seidengazefabrik AG
Zürich
Grütlistr. 68
Ch 78027 Zürich 2
Tel.:01. 256825
FAX:01. 2010187

Nylon gaze

SERVA FEINBIOCHEMICA
GmbH & Co. KG
P. O. Box 10 52 60
Carl-Benz-Str. 7
W 6900 Heidelberg 1
Tel.:0049. 6221. 502-0
FAX:0049. 6221. 502-188

Biochemicals

Suppliers	**Products**

SIGMA Chemie GmbH
Diagnostic
Grünwalder Weg 30
W 8024 Deisenhofen
Tel.:   0049. 89. 61301-0
          0130. 5155
FAX: 0049. 89. 6135135
          0130. 6490
	Biochemicals and antibodies

SIGMA CHEMICAL CO. P. O. Box 14508 St. Louis MO 63178, USA Tel.:001. 314. 771-5765 FAX:001. 800.325-5052	s.a.
Southern Biotechnology Associates INC. P. O. Box 26221 Birmingham AL 35226, USA Tel.:001. 205. 945-1774 FAX:001. 205. 945-8768	Dyes and antibodies
Southern Biotechnology Associates INC. **Germany → DUNN Labortechnik**	
Spectra-Physics 1250 W. Middlefield Rd. Mountain View, CA 94039, USA Tel.:001. 415. 961-2550	Laser manufacturers: Argon and krypton laser CW dye lasers He-Ne laser
Spindler & Hoyer GmbH & Co. Postfach 3353 Königsallee 23 W 3400 Göttingen Tel.:0049. 551. 6935-0 FAX:0049. 551. 6935-166	Optical supply: Mini-bench series of small components and mounts Laser manufacturers: Yellow,orange, red He-Ne laser
VERITY Software House, Inc. P. O. Box 247 Topsham - Maine 04086, USA Tel.:001. 207. 729-6767	Software products for the PC: IsoContour, 2-parameter isometric contour generator ModFit, DNA histogram modeling program
Whatman International Ltd. Maidstone, England	Paper for cleaning optical equipment
Wild Leitz Postfach 2020 W 6330 Wetzlar Tel.:0049. 6441. 29-0 FAX:0049. 6441. 293399	Microscopes
Yakult Honsha Co. Ltd. 1-1-19 Higashi Shinbashi Minato-ku Tokyo, 105, Japan	Enzymes and biochemicals

Zeiss, Carl
Postfach 1380
Carl-Zeiss-Str.
W 7082 Oberkochen
Tel.:0049. 7364. 20-0
FAX:0049. 7364. 20-6808

Optical filters, prisms, etc.

Zynaxis Cell Science INC.
371 Phoenixville Pike
Malvern PA 19355, USA
Tel.:001. 215. 889-2200
FAX:001. 215. 889-2222

Chemicals